WORLD BANK TECHNICAL PAPER NUMBER 123

Irrigation in Sub-Saharan Africa

The Development of Public and Private Systems

Shawki Barghouti and Guy Le Moigne

The World Bank
Washington, D.C.

Copyright © 1990
The International Bank for Reconstruction
and Development/THE WORLD BANK
1818 H Street, N.W.
Washington, D.C. 20433, U.S.A.

All rights reserved
Manufactured in the United States of America
First printing May 1990

Technical Papers are published to communicate the results of the Bank's work to the development community with the least possible delay. The typescript of this paper therefore has not been prepared in accordance with the procedures appropriate to formal printed texts, and the World Bank accepts no responsibility for errors.

The findings, interpretations, and conclusions expressed in this paper are entirely those of the author(s) and should not be attributed in any manner to the World Bank, to its affiliated organizations, or to members of its Board of Executive Directors or the countries they represent. The World Bank does not guarantee the accuracy of the data included in this publication and accepts no responsibility whatsoever for any consequence of their use. Any maps that accompany the text have been prepared solely for the convenience of readers; the designations and presentation of material in them do not imply the expression of any opinion whatsoever on the part of the World Bank, its affiliates, or its Board or member countries concerning the legal status of any country, territory, city, or area or of the authorities thereof or concerning the delimitation of its boundaries or its national affiliation.

The material in this publication is copyrighted. Requests for permission to reproduce portions of it should be sent to Director, Publications Department, at the address shown in the copyright notice above. The World Bank encourages dissemination of its work and will normally give permission promptly and, when the reproduction is for noncommercial purposes, without asking a fee. Permission to photocopy portions for classroom use is not required, though notification of such use having been made will be appreciated.

The complete backlist of publications from the World Bank is shown in the annual *Index of Publications*, which contains an alphabetical title list (with full ordering information) and indexes of subjects, authors, and countries and regions. The latest edition is available free of charge from the Publications Sales Unit, Department F, The World Bank, 1818 H Street, N.W., Washington, D.C. 20433, U.S.A., or from Publications, The World Bank, 66, avenue d'Iéna, 75116 Paris, France.

ISSN: 0253-7494

Shawki Barghouti is division chief of the Agriculture Production and Services division of the Agriculture and Rural Development department of the World Bank. Guy Le Moigne is senior adviser, Agriculture and Water Resources, Agriculture and Rural Development department.

Library of Congress Cataloging-in-Publication Data

Barghouti, Shawki M.
   Irrigation in sub-Saharan Africa : the development of public and
private systems / Shawki Barghouti and Guy Le Moigne.
     p. cm. — (World Bank technical paper, ISSN 0253-7494 ;
no. 123)
   Includes bibliographical references.
   ISBN 0-8213-1554-4
   1. Irrigation—Africa, Sub-Saharan. 2. Irrigation farming–
–Africa, Sub-Saharan. I. Le Moigne, Guy, J.–M., 1932–   II. Title.
III. Series.
TC919.S73B37 1990
333.91'3'0967—dc20   90-36097
                                                                     CIP

FOREWORD

One of the major economic concerns in the world today is how to increase agricultural production in Sub-Saharan Africa (SSA), both in the short term and on a long term, sustainable basis. Since the mid-1970s, the story in SSA agriculture has been one of long term decline in agricultural production, cycles of drought, increasing population pressure on the land, a loss of export markets and an alarming rise in food imports. The framework of a strategy for reversing these trends has been outlined in the recent World Bank study entitled "Sub-Saharan Africa: From Crisis to Sustainable Growth - A Long-Term Perspective Study".

Many African governments believe that large-scale irrigation would enable their countries to reach food self-sufficiency and even allow them to earn foreign exchange from the export of crops. The World Bank's experience with irrigation in Africa has been mixed, however, with frequent project cost overruns, lower economic rates of return than expected, and institutional problems which inhibit operations and maintenance. This publication, prepared as a joint effort between the staff of the Africa Region and the Agricultural and Rural Development Department of the World Bank, reviews the past experience with irrigation projects and strategies, examines the reasons for past problems and identifies past successes. It also explores the potential for future irrigation development in Africa and suggests strategies and techniques for realizing that potential.

This publication should be of interest to anyone involved in agricultural development in Africa. Its review of past failures should help to avoid repeating the mistakes of the past, and the stories of successful cases provide models which may be applicable to other situations. It analyzes options for the development of irrigation, both at the technical and at the policy levels.

The Bank has been stressing the importance of a favorable macro-economic environment as a precondition for successful development in Africa since 1980, and a large number of African countries have been responding to this call for action by undertaking structural adjustment programs in an attempt to improve their economic performance. It is now important that these efforts be supported by sector programs and investment projects which exploit the opportunities offered by these more favorable conditions in order to stimulate African agriculture to more rapid expansion than has been the case during the 1970s and 80s. We hope that this document will provide useful insights and guidance for practical means to realize this goal and strongly recommend it as background information for those working in this field.

Two important messages which emerge from the studies are that small scale private irrigation has a much better record of success than large scale public irrigation schemes, and that the public sector should play a very important role in the planning, regulation, and management of water resources. We believe that this is an important and valuable conclusion, and suggest that it has significant implications for future strategies and project designs.

Hans Wyss
Director
Africa Technical Department

Michel J. Petit
Director
Agriculture and Rural Development Department

TABLE OF CONTENTS

EXECUTIVE SUMMARY vii

PART I – IS IRRIGATION IN AFRICA A GOOD INVESTMENT? THE PROS AND CONS

CHAPTER 1: PROSPECTS FOR IRRIGATION DEVELOPMENT IN SUB-SAHARAN AFRICA 1
by Hans Wyss
- Small-Scale Irrigation 1
- The Private Sector In Irrigation Development 2
- What Does The Future Hold? 4

CHAPTER 2: THE POTENTIAL FOR IRRIGATION DEVELOPMENT IN SUB-SAHARAN AFRICA 5
by Jose Olivares
- The Irrigation Potential Of Sub-Saharan Africa 5
- The Cost Of Developing Irrigation In SSA 10
- Conclusion 13

CHAPTER 3: PRIVATE SECTOR IRRIGATION IN AFRICA 17
by David Seckler
- The Synergism Between Irrigation And Other Factors Of Production ... 18
- Irrigation Needs And Potential In Africa 21
- Water Deficiency Factors 22
- Irrigation Cultures, Technologies, And Institutions 24
- Pump Irrigation Systems 25
- Engineering And Economic Aspects Of Pump Irrigation Systems 26
- The Case Of Senegal 27
- The Case Of Botswana 30
- The Case Of Nigeria And Mali 31
- Tank Irrigation Systems 31
- Intermediary Agencies For Private Sector Irrigation In Africa ... 34
- Irrigation Policy And Evaluation 36

CHAPTER 4: HOW RISKY IS IRRIGATION DEVELOPMENT IN SUB-SAHARAN AFRICA? 45
by Guy Le Moigne and Shawki Barghouti
- A Review Of Irrigation In Sub-Saharan Africa 45
- Erratic Rainfall Regimes 47
- Constraints To Irrigation In SSA 52
- Managing Common Water Resources For Irrigation And Water Supply .. 55
- Farmer Participation In Water Management 55
- The Environmental Dimension 56
- Recommendations For Developing Irrigation In SSA 57

CHAPTER 5: SECTORAL STRATEGY FOR IRRIGATION DEVELOPMENT IN SUB-SAHARAN AFRICA: SOME LESSONS FROM EXPERIENCE 60
by Uma Lele and Ashok Subramanian
- Introduction 60
- Irrigation In Nigeria And Kenya 61
- Lessons from Experience 65

PART II — THE LESSONS

CHAPTER 6: TECHNICAL ISSUES OF IRRIGATION DEVELOPMENT IN SUB-SAHARAN AFRICA 69
by Akhtar Elahi and B. Khushalani
- Reasons For Choosing Irrigation 69
- Points To Consider Before Investing In Irrigation Projects 72
- A Review Of Three Irrigation Projects In SSA 75
  - (1) Sudan: Rahad Irrigation Project. 75
  - (2) Madagascar: Morondava Irrigation and Rural Development Project. 77
  - (3) Madagascar: Lak Alaotra Irrigation Project. 80
- Conclusion 82

CHAPTER 7: POLICY AND MANAGEMENT PROBLEMS 83
by Shawki Barghouti and Ashok Subramanian
- Policy Environment 83
- Fiscal Policy 84
- Project Management and Performance 85
- Conclusion 87

CHAPTER 8: COUNTRY CASE STUDIES 88

A. PRIVATE SECTOR IRRIGATION IN ZIMBABWE 88
by Johannes ter Vrugt
- Present Situation 88
- Water Supplies 89
- Irrigation Systems 89
- Types Of Irrigation Development 89
- Investment Costs 90
- Irrigated Crops 90
- The Future 90

B. PRIVATE IRRIGATION DEVELOPMENT IN THE SENEGAL VALLEY 91
by Salah Darghouth
- Background 91
- Present Irrigation Schemes In The Senegal Valley 91
- Constraints On The Development Of Publicly Funded Irrigation 92
- Recent Expansion Of Privately Funded Irrigation In Mauritania 93
- Prospects For Future Irrigation Development In Mauritania 94

C. PRIVATE SECTOR SMALL-SCALE IRRIGATION IN NIGERIA 95
by Lewis G. Campbell
- Irrigation Development 95
- Present Situation 97
- Issues Requiring Attention 97
- Conclusion 99

EXECUTIVE SUMMARY

The seven papers which are included in this publication have been prepared in an attempt to provide background information and guidance to those who are designing agricultural strategies and irrigation investment projects for Sub-Saharan African (SSA) countries. Specifically, it seeks to clarify the role which irrigated agriculture might play in these strategies and how that role can be effectively realized. It reviews problems which have arisen in past attempts to expand irrigated agriculture in SSA as well as examples of successful irrigation development, and proposes considerations for future design and implementation based on the lessons of the past.

The Physical Potential for Irrigation Development

Several of the papers stress the potential importance of irrigated agriculture in a continent where rainfall is frequently marginal and erratic, or confined to a limited portion of each year. The total physical potential for expanding irrigation in SSA is examined in Chapters 2, 3 and 4 with the conclusion that the 5 million hectares presently under irrigation, equal to 2 percent of the presently cropped area in SSA, could eventually be expanded to about 20-25 million hectares. While this potential is less per capita than in Asia or Latin America, it nevertheless represents a substantial potential increase in agricultural output since yields on irrigated land average three and one half times those from rainfed farming. The countries with the largest physical potential for expansion are Angola, Zaire, Sudan, Mozambique, Tanzania, Nigeria, the Central African Republic and Chad, according to a 1985 FAO study, although irrigation could play an important subsidiary role in many other SSA countries.

Past Problems in Irrigation

A number of the papers describe the problems encountered in past irrigation strategies and projects. While about 75 percent of all SSA irrigation projects reviewed in a recent Bank study achieved or exceeded the expected rate of economic return, some major projects failed.

Cost Overruns are a fairly frequent cause of reduced rates of economic return. The commonly held view that irrigation installation costs per hectare are higher in SSA than in other areas of the world is examined in several of the papers; the conclusion is that with adequate planning and careful design, costs should be no higher in Africa than elsewhere.

Institutional factors are cited as the basis for deteriorating maintenance and declining output from large scale public irrigated schemes. The widespread decline in the effectiveness of public institutions in Africa has had its effect on these systems as well.

The policy environment for irrigation is cited in Chapters 3 and 7 as an important determinant of the success or failure of individual projects, including the availability of inputs, adequate budgetary resources, accessible markets, reasonable producer prices, and an adequate physical infrastructure. These conditions frequently are not met in SSA, although conditions may be improving in countries undertaking structural adjustment reforms.

<u>Technical design issues</u> can also lead to failures if improper choices of location or design are made.

<u>Cultural factors</u> are cited in several of the studies as an important determinant of project success or failure, although several instances of successful transitions from pastoral communities to irrigated farming are mentioned.

<u>Environmental problems</u> resulting from expanded irrigation are described in Chapter 4. These include the increased incidence of disease, especially schistosomiasis and malaria, and the reduction in size of floodplains and swamps with its impact on fishing, wildlife and cultivation at the edges of these areas.

Past Successes in Irrigation

A pervasive theme which emerges from a number of the papers is the relatively successful record of small scale private irrigation systems in contrast to the spotty record of large scale public schemes. Chapters 1 and 3 elaborate on the reasons for this difference, and Chapter 8 presents interesting examples of successful small scale irrigation strategies in three SSA countries.

Several of the papers point out that small scale irrigation has certain design and institutional limitations and may not always be the correct solution in every case. But the evidence is clear that where applicable, the chances of successful irrigation development are greatly enhanced when small scale private systems can be used.

Suggestions for How to Proceed

Chapter 1 outlines twelve general factors to take into account in the design of private sector irrigation projects, including choice of technology, defining the role of government and the possible need for regulation, institutional arrangements, land tenure, and the general macro-economic setting.

Chapter 3 provides information regarding the basis of choice for selecting alternate systems for water supply, conveyance, and field application, concluding that privately owned pumps and small scale tank irrigation systems are usually the most appropriate for SSA conditions. Chapter 3 also suggests a "growth centers" strategy where irrigated areas with high potential and reasonable infrastructure could be linked to other more backward areas in an expanding network.

Chapters 4, 5, 6 and 7 make recommendations regarding the measures that need to be taken when expanding investment in irrigation in order to improve the prospects for success. Chapter 4 stresses the need to improve the data base, rehabilitate existing systems, conduct research and carry out pilot projects to test new technologies, improve cultivation practices, put more emphasis on planning and monitoring water studies, and take into account environmental and social factors. Chapter 5 stresses the need for a carefully formulated irrigation subsector strategy and the need to develop institutional capacity for analyzing, formulating and managing the subsector programs.

Chapter 6 stresses that irrigation schemes should not be too large for the development environment, the importance of an integrated approach, keeping capital costs at a reasonable level, cost recovery policies conceived and agreed in advance, and technology appropriate for the African scene. Chapter 7 stresses the need to assure both a favorable policy environment and careful project preparation and management.

Summary

Taken together, the papers make a strong case for the important role which irrigated agriculture can and should play in SSA's future development. The actual possibilities will vary considerably by country, and careful project choices and designs will be needed to avoid the mistakes of the past. But each country agricultural strategy should consider what role irrigated agriculture can play in a given situation, with special emphasis on the prospects for small scale private irrigation systems where the best results have been achieved in the past.

PART I

IS IRRIGATION IN AFRICA A GOOD INVESTMENT?

THE PROS AND CONS

CHAPTER 1: PROSPECTS FOR IRRIGATION DEVELOPMENT IN SUB-SAHARAN AFRICA

Hans Wyss
(Director, Africa Technical Department, World Bank)

Every spring the fear of drought looms large in much of Africa. The human misery it has already caused there is staggering. Large parts of the continent are arid or semiarid and the farmers have little hope of obtaining good yields without some type of irrigation. Although irrigation is now practiced on about 14.5 percent of the world's arable land, the figure is only about 2 percent in Africa (compared with 29 percent in Asia). Nonetheless, some people believe that irrigation is a luxury and not a necessity in this region.

One reason is that Africa's experience with irrigation projects has been mixed. A recent report on irrigation in the region indicates that about 75 percent of all projects reviewed by the authors have achieved or exceeded the expected rate of economic return, and that the costs of construction were not much different from those in other parts of the world. At the same time, some major projects have failed. In addition, the benefits in other cases have fallen short of project goals, especially with respect to yields. Also the areas already brought under irrigation have been considerably smaller than planned, although capital investments have reached US$25,000 per hectare (and even US$45,000 in Ghana) on some projects. These results have made many investors skeptical about the wisdom of starting new irrigation projects in Africa.

Small-Scale Irrigation

The performance of large irrigation schemes that have been developed, operated, and maintained by government agencies in Africa has been the target of criticism for some time now. I need not dwell on the various factors that are making large irrigation projects so difficult to implement and manage there. Instead, I want to focus on medium- and small-scale irrigation, especially systems owned and operated by private enterprises, as therein lies the hope for increasing food production in Africa at present.

A number of countries are beginning to allow the private sector to play a substantial role in the development, operation, and maintenance of irrigation schemes, with minimal support from governments. In fact, this is not a new approach in Africa. For centuries African communities have been practicing irrigation wherever water is available: in coastal swamps and estuaries along the west coast of Africa from the Casamance River to the Zaire River; in inland swamps and floodplains along the Senegal, Niger, and Bani rivers; in the lowlands in Sierra Leone, Ghana, and Ivory Coast; around Lake Victoria and Lake Malawi; and along small streams in Madagascar, Tanzania, and Zimbabwe. Furrow irrigation from rivers has been practiced throughout the continent and pumping by hand using the traditional shaduf or similar mechanical devices is also common. Small pumps are purchased by farmers on their own initiative or through government-aided programs to exploit shallow groundwater.

Private small-scale irrigation generally develops spontaneously when farmers wish to intensify food production in response to rising population pressure, land demands, and market opportunities. Of the estimated 5.3 million hectares of irrigated land in Africa, 2.7 million ha are accounted for by private flood, swamp, surface, and low-lift irrigation developed largely

by small farmers without government support. Half a million ha have been developed with modern irrigation facilities by private sector commercial farmers. In sum, about 60 percent of the irrigated land receives no government funding, and this figure is increasing.
However, the productive capacity of the traditional small private farmer is severely limited both in terms of area under crop each year and low crop yields due to inefficient and outmoded irrigation techniques and calls for ways to improve the system for increased productivity.

The private schemes may be operated by one owner, a community, a group of tenants, or a commercial estate. They produce food crops, industrial crops, fruit, vegetables, and fodder for livestock. Private small- or medium-scale irrigation tends to cost less per hectare than (public) large schemes. The contribution of local labor or resources to the construction, operation, and maintenance costs relieves the government of much of the investment burden. Also, the economic value of these contributions is often considerable, given the strong competition that exists for privately employed labor and other resources. The execution of irrigation works has also given small local contractors an opportunity to prove themselves. Thus it is fair to say that the record of success for private irrigation has been quite good in Africa, and considerably better than that for the large state-operated projects. However, private schemes do have their limitations in terms of hectarage, reliability, possibility of intensification, and level of yields.

The Private Sector in Irrigation Development

Although existing public investments in irrigation in Africa need to be improved, more emphasis should be put on private sector irrigation. Small-scale irrigation holds the promise of considerable productivity gains. The countries of Africa must capitalize on this potential, particularly if they are to meet the needs of their growing populations. Although the risks involved are by no means small, the public sector can help to soften these risks--by providing credit, subsidized energy, and the proper setting or investment climate in agriculture. Indeed, many governments have already made it possible for the private sector to invest in profitable irrigation technologies. Before any further action is taken, however, it is vital for all those concerned with the possibility of expanding irrigation in Africa to lay down some general principles that will help to ensure the success of future projects in this region. Some of these principles have already emerged from this workshop. They can be categorized under 12 broad headings.

Choice of technology. The ideal irrigation technology for developing countries is one that will be profitable, but also technically simple, so that systems can easily be constructed and maintained at the village level, as in the case of the private tubewells that have sprung up throughout Southeast Asia. The success of sprinkler irrigation on private farms in Zimbabwe suggests that it might also be possible to make the jump to more advanced technologies in some parts of Africa. Although the Zimbabwe example is not typical for Africa, it shows that modern irrigation and modern production techniques can be profitable in this region and merit closer consideration. In any case, one must remember that irrigation technology does not function alone, but in concert with other inputs, and that the choice is ultimately governed by the available infrastructure and the level of agricultural technology in the country.

Hydrological data. Investment must not proceed without a thorough understanding of local hydrology, soils, and topography. Although this may sound axiomatic, in some cases managers have found themselves burdened by cost overruns and environmental problems for this very reason.

Regulation. Opinion is divided on the need for regulation—if enforced, regulations frequently work against the poorer farmers; if not, mayhem may follow. No one would deny, however, that it is vital to monitor and regulate the abstraction rates of water resources, particularly where the risk of depletion is high.

The Government as a catalytic agent. No irrigation system can be expected to work if farmers fail to adopt it. Therefore the government may have to start the ball rolling, with a view to eventually relinquishing its catalytic role and encouraging the private sector to take over. The government could function as a commercial entity, for example, so that the private sector does not face a competitive disadvantage as it tries to take the reins of individual projects. Whatever strategy is used, a clear government commitment to private sector development is essential. That is to say, government policies must help to shape this strategy.

Water user associations. Irrigation will be difficult to organize and manage without some involvement of the users. The appropriate institutional structure may turn out to be a village association, a cooperative, or some other type of system. Needless to say, the social environment will have to be examined to determine what type of structure will best match local customs and conditions.

Conditions for successful private sector irrigation investments. Obviously, many factors contribute to the success of private sector irrigation investments—including legal access to an unappropriated resource, strict enforcement of legal rights, institutional development of the domestic capital market, socioeconomic conditions favoring cooperative investments, legal protection for private investments, relatively low cost of capital, a high ratio of rural labor to arable land, an infrastructure that permits relatively low-cost access to larger markets, and policies that do not shift the internal terms of trade against agriculture. It is important to define in advance which of these factors must be present before a project is undertaken. Although many of these factors will have direct bearing on the "profitability" of a project, one must not forget that profitability also extends to social benefits. One of the large challenges in Africa is to get the private sector to reflect social costs and benefits by investing in socially desirable activities. International organizations can help protect the private sector from noncommercial risks. It will also be important to help the private sector gain access to much-needed foreign currency and to obtain local credit for medium-term investment.

Sustainability. Remember that incentives are required to ensure sustainability—for example, via tax policies. Also, attention must be given to the long-term impact of irrigation systems, particularly pumping from aquifers, which could have disastrous results if the resource was depleted. Although the impact of individual tubewell development projects has been studied extensively, little work has been done on the total effects of large-scale private sector tubewell development in a given area. Another question to consider is what would happen to areas beyond those where private sector development was occurring. Would the public sector be responsible for development there as well?

Equity. One criticism of private sector irrigation is that it can have an adverse effect on equity over the long term, as can be seen from irrigation development in India and in Mauritania, where development has tended to exacerbate the differences between the rich and the poor. However, the undeveloped potential for irrigation is so vast in Africa at present that equity is not a major concern at this juncture.

Land tenure. The impact of tenure on irrigation varies from one country to another. In some cases (e.g., Sudan), it does not appear to have constrained the growth of irrigation, whereas in others (the communal areas of Zimbabwe and Somalia) tenure has been a serious issue. Consequently, the question of tenure must be evaluated on a case-by-case basis in this region.

Pricing. Although successful irrigation schemes in Asia have all had some form of price support, subsidies might merely encourage inefficient irrigation in Africa. Perhaps what is more important is price stabilization, especially in semiarid areas, which desperately need to have stable incomes and yields.

The capital constraint. Without adequate capital, no country will be able to adopt the irrigation technologies it envisions. Credit was the vital element leading to the adoption of tubewells in Asia, and to the use of small pumps in Nigeria, and to successful private sector development in Morocco. The important question to ask here is what credit system would be most effective, especially in countries whose financial structure is just evolving.

Organizational options. Various organizational arrangements can be considered for implementing private sector irrigation. One innovative idea is to rely on intermediary institutions for this purpose, such as private voluntary organizations, cooperatives, public utility companies, and agricultural development organizations, which could work between government agencies and private enterprises to promote private sector irrigation. Specialized institutions are also needed to monitor resource use, handle environmental questions, and oversee infrastructure requirements.

What Does The Future Hold?

It is now widely believed that selective development of irrigation could have beneficial effects in a number of African countries. Unfortunately, many of these countries are not in a financial position to move ahead with a program of irrigation development, and some already have sunk costs in irrigation. These are substantial and cannot be ignored. One way to help them get irrigation development going would be for agencies such as the World Bank to consider a broader range of economic factors such as reduced risk and increased flexibility of farming systems in calculating an acceptable rate of return, particularly in view of the special problems associated with irrigation development in Africa. Another possibility would be to revive the old-fashioned irrigation sector loan. Such loans could provide support for private investments, both for improving sunk cost schemes and for new operations. A selective approach should obviously be used in applying this idea to individual countries.

These and other recommendations are now on the table and ready for in-depth analysis. The countries of Africa cannot afford to wait much longer for a solution to their enormous agricultural problems. Private sector irrigation is without doubt the logical strategy for these countries to follow if they wish to see a brighter economic future.

CHAPTER 2: THE POTENTIAL FOR IRRIGATION DEVELOPMENT IN SUB-SAHARAN AFRICA

Jose Olivares
(Principal Agricultural Economist, Agriculture Operations Division, Southern Africa Department, World Bank)

Irrigation development, together with new varieties, extension services, and fertilizer supplies, are among the most important factors behind the impressive agricultural growth experience in recent years in Southeast Asia. Consequently, many countries in Africa have been wondering whether irrigation could play a similar role in the development of their agriculture. The two most relevant questions in this regard are (1) How much of Africa could eventually be irrigated? and (2) What would it cost to develop that irrigation potential? This chapter examines these questions in the light of information from recent studies.

The Irrigation Potential Of Sub-Saharan Africa

Two estimates of the irrigation potential of Sub-Saharan Africa (SSA) are available. One, prepared by the U.N. Food and Agriculture Organization (FAO), covers the entire continent; the other, prepared by the World Bank with UNDP support, covers six countries.[1] In this discussion, the more specific results of the latter are used to reassess the more general results of the former. This exercise produces a more realistic assessment of the likely irrigation potential of the region.

Africa south of the Sahara irrigates some 5 million hectares. The irrigated area has been growing at the rate of about 150,000 hectares per year since the mid-1960s. This is equivalent to 5 percent a year in 1965-74 and less than 4 percent a year in 1974-82. Seventy percent of the total irrigated area lies in three countries: Sudan (1,750,000 hectares, 35 percent, Madagascar (960,000 hectares, 20 percent), and Nigeria (850,000 hectares, 17 percent). Four other countries--Mali, Tanzania, Zimbabwe, and Senegal, which irrigate between 100,000 and 160,000 hectares--cover another 530,000 hectares, or 10 percent of the irrigated area (Table 2.1). The governments of these various countries have developed 2.1 million hectares, primarily under large modern schemes; traditional methods have been used on 2.4 million hectares, and the remaining 0.5 million hectares have been developed by the modern private sector. According to FAO estimates, these 5 million hectares contribute 10 percent (5.3 million tons) to regional cereal supplies, and 6 to 8 percent to root and vegetable supplies. The total irrigation potential of SSA is about 33.6 million hectares (Table 2.2). Although this seems to be quite a small area (some absolute and relative comparisons with the actual or potential irrigated area of other regions are presented below), I believe the irrigation potential is actually smaller.

---

[1] FAO, *Irrigation in Africa South of the Sahara*, Investment Centre Technical Paper 5 (Rome, 1986); J. Olivares, *Options and Investment Priorities in Irrigation Development*, World Bank, Agriculture and Rural Development Department Report (Washington, D.C., August 1987); and individual country reports on Mali and Sudan (1985), Botswana (1986), and Zambia, Kenya, and Zimbabwe (1987). The figures provided in the text are based on these reports, except where otherwise noted.

Table 2.1: IRRIGATION IN AFRICA RANKED BY IRRIGATED AREA IN 1982

COUNTRY	IRRIGATION POTENTIAL '000 ha	% OF LAND AREA	ESTIMATED IRRIGATED AREA ('000 ha) 1965	1974	1982	IRRIGATED /TOTAL CROP (1982)	REMAINING POTENTIAL AFTER 1982 ABSOLUTE	% OF POTENTIAL
Sudan	3,300	1.3	1352.0	2064.0	1750	14.1	1550	47.0
Madagascar	1,200	2.1	330.0	426.0	960	32.0	240	20.0
Nigeria	2,000	2.2	300.0	400.0	850	2.8	1150	57.5
Mali	340	0.3	117.0	150.3	160	7.8	180	52.9
Tanzania	2,300	2.6	27.9	52.0	140	2.7	2160	93.9
Zimbabwe	280	0.7	36.0	65.4	130	4.6	150	53.6
Senegal	180	0.9	13.6	38.1	100	1.9	80	44.4
Ethiopia	670	0.6	58.7	92.6	87	0.6	583	87.0
Somalia	87	0.1	31.5	47.2	80	7.2	7	8.0
Mozambique	2,400	3.1	16.0	34.0	70	2.3	2330	97.1
Swaziland	7	0.4	20.0	30.0	60	43.5	-53	-757.1
Sierra Leone	100	1.4	N.A.	N.A.	55	3.1	45	45.0
Ivory Coast	130	0.4	6.7	32.9	52	1.3	78	60.0
Burundi	52	2.1	0.0	5.5	52	4.3	0	0.0
Chad	1,200	0.9	7.0	10.0	50	1.9	1150	95.8
Kenya	350	0.6	15.0	36.6	49	2.1	301	86.0
Guinea	150	0.6	41.0	64.0	45	2.9	105	70.0
Niger	100	0.1	1.9	2.9	30	0.9	70	70.0
Burkina Faso	350	1.3	0.1	6.0	29	1.1	321	91.7
Gambia	72	6.6	25.0	33.2	26	16.3	46	63.9
Zaire	4,000	1.8	2.5	4.2	24	0.4	3976	99.4
Mauritania	39	0.1	10.7	33.0	23	11.1	16	41.0
Benin	86	0.8	0.1	9.4	22	1.2	64	74.4
Cameroon	240	0.5	4.0	14.2	20	0.3	220	91.7
Malawi	290	3.1	0.6	11.0	20	0.9	270	93.1
Liberia	N.A.	N.A.	0.0	2.5	19	5.2	N.A.	N.A.
Zambia	3,500	4.7	2.0	17.5	16	0.3	3484	99.5
Rwanda	44	1.8	2.6	12.7	15	1.5	29	65.9
Mauritius	N.A.	N.A.	12.0	21.0	14	13.1	N.A.	N.A.
Togo	86	1.5	0.2	6.2	13	0.9	73	84.9
Uganda	410	2.1	3.0	16.0	12	0.2	398	97.1
Botswana	100	0.1	3.0	9.5	12	0.9	88	88.0
Ghana	120	0.5	0.2	16.1	10	0.4	110	91.7
Angola	6,700	5.3	N.A.	N.A.	10	0.3	6690	99.9
Congo	340	1.0	0.0	2.2	8	1.2	332	97.6
Central African Republic	1,900	3.0	N.A.	N.A.	4	0.2	1896	99.8
Gabon	440	1.7	0.0	0.2	1	0.2	439	99.8
Lesotho	8	0.3	0.0	0.0	1	0.4	7	87.5
Guinea Bissau	70	1.9	N.A.	N.A.	N.A.	N.A.	N.A.	N.A.
Equatorial Guinea	N.A.		N.A.	N.A.	N.A.		N.A.	
TOTAL	33,641		2440.3	3766.4	5019		28585	

Source: FAO, Irrigation in Africa South of the Sahara, Investment Center, Technical Paper No. 5 (Rome 1986).

Table 2.2: IRRIGATION IN AFRICA RANKED BY IRRIGATION POTENTIAL

COUNTRY	IRRIGATION POTENTIAL '000 ha	% OF LAND AREA	ESTIMATED IRRIGATED AREA ('000 ha) 1965	1974	1982	IRRIGATED /TOTAL CROP (1982)	REMAINING POTENTIAL AFTER 1982 ABSOLUTE	% OF POTENTIAL
Angola	6,700	5.3	N.A.	N.A.	10	0.3	6690	99.9
Zaire	4,000	1.8	2.5	4.2	24	0.4	3976	99.4
Zambia	3,500	4.7	2.0	17.5	16	0.3	3484	99.5
Sudan	3,300	1.3	1352.0	2064.0	1750	14.1	1550	47.0
Mozambique	2,400	3.1	16.0	34.0	70	2.3	2330	97.1
Tanzania	2,300	2.6	27.9	52.0	140	2.7	2160	93.9
Nigeria	2,000	2.2	300.0	400.0	850	2.8	1150	57.5
Central African Rep.	1,900	3.0	N.A.	N.A.	4	0.2	1896	99.8
Madagascar	1,200	2.1	330.0	426.0	960	32.0	240	20.0
Chad	1,200	0.9	7.0	10.0	50	1.9	1150	95.8
Ethiopia	670	0.6	58.7	92.6	87	0.6	583	87.0
Gabon	440	1.7	0.0	0.2	1	0.2	439	99.8
Uganda	410	2.1	3.0	16.0	12	0.2	398	97.1
Burkina Faso	350	1.3	0.1	6.0	29	1.1	321	91.7
Kenya	350	0.6	15.0	36.6	49	2.1	301	86.0
Mali	340	0.3	117.0	150.3	160	7.8	180	52.9
Congo	340	1.0	0.0	2.2	8	1.2	332	97.6
Malawi	290	3.1	0.6	11.0	20	0.9	270	93.1
Zimbabwe	280	0.7	36.0	65.4	130	4.6	150	53.6
Cameroon	240	0.5	4.0	14.2	20	0.3	220	91.7
Senegal	180	0.9	13.6	38.1	100	1.9	80	44.4
Guinea	150	0.6	41.0	64.0	45	2.9	105	70.0
Ivory Coast	130	0.4	6.7	32.9	52	1.3	78	60.0
Ghana	120	0.5	0.2	16.1	10	0.4	110	91.7
Botswana	100	0.1	3.0	9.5	12	0.9	88	88.0
Sierra Leone	100	1.4	N.A.	N.A.	55	3.1	45	45.0
Niger	100	0.1	1.9	2.9	30	0.9	70	70.0
Somalia	87	0.1	31.5	47.2	80	7.2	7	8.0
Benin	86	0.8	0.1	9.4	22	1.2	64	74.4
Togo	86	1.5	0.2	6.2	13	0.9	73	84.9
Gambia	72	6.6	25.0	33.2	26	16.3	46	63.9
Guinea Bissau	70	1.9	N.A.	N.A.	N.A.	N.A.	N.A.	N.A.
Burundi	52	2.1	0.0	5.5	52	4.3	0	0.0
Rwanda	44	1.8	2.6	12.7	15	1.5	29	65.9
Mauritania	39	0.1	10.7	33.0	23	11.1	16	41.0
Lesotho	8	0.3	0.0	0.0	1	0.4	7	87.5
Swaziland	7	0.4	20.0	30.0	60	43.5	-53	-757.1
Mauritius	N.A.	N.A.	12.0	21.0	14	13.1	N.A.	N.A.
Equatorial Guinea	N.A.	N.A.	N.A.	N.A.	N.A.	N.A.	N.A.	N.A.
Liberia	N.A.	N.A.	0.0	2.5	19	5.2	N.A.	N.A.
TOTAL	33,641		2440.3	3766.4	5,019		28,585	

Source: FAO, Irrigation in Africa South of the Sahara, Investment Centre, Technical Paper No. 5 (Rome, 1986).

FAO estimates tend to have an upward bias, in part because they are based on the suitable soils located in reasonable proximity to adequate surface runoff from hills and mountains--or in alluvial areas with substantial groundwater recharge--in zones that have the same length of growing period. Rather high efficiencies were assumed for the exercise, and the growing period zones may cut across watersheds. This methodology is adequate for a continent-wide reconnaissance estimate, but too general to provide operationally meaningful results.

Also, the scale at which the estimates were produced (1:500,000) is too large to give significant results at the country level. Moreover, since the FAO works from estimates of water availability, its figures on irrigation potential are positively correlated with rainfall. Eight of the 10 countries said to have the largest potential (1.2 million hectares or more) receive substantial rainfall in much of their territory. These countries are Angola, Zaire, Zambia, Mozambique, Tanzania, Nigeria, the Central African Republic, and Madagascar (see Table 2.2). Although they may have "technical" potential--in the sense of enough land and, obviously, enough water--irrigation might prove unnecessary in many of these areas precisely because rainfall is adequate. The total irrigation potential estimated by the FAO for these 8 countries is 24 million hectares, which is more than 70 percent of the estimate for the entire region.

The World Bank has developed a methodology for producing more detailed assessments of irrigation potential in individual countries, although still at a reconnaissance level. The methodology was successfully tested in 10 countries, 6 of which (Mali, Sudan, Botswana, Kenya, Zambia, and Zimbabwe) are in SSA. In this case, the estimates of irrigation potential were prepared on a project-by-project basis. Thus, both the waters and the soils to be combined under irrigation are in the same watershed, and both can be physically linked through engineering facilities. These estimates--although they cost more and take longer to produce--are closer to each country's "true" potential.

For five of the six SSA countries studied (excluding Zambia, for reasons explained below), the irrigation potential is estimated to be 3.7 million hectares. This figure is 15 percent less than the 4.4 million hectares estimated by the FAO for the same countries (Table 2.3). If the same proportion (85 percent) were to hold for the continent at large [2] a better estimate of the irrigation potential of the region would be about 28.5 million hectares.

Zambia may be an extreme case. The FAO estimate is 3.5 million hectares, whereas we put it at 423,000 hectares, on the optimistic assumption that the electric company would surrender its rights on the Kafue River waters. Thus, the FAO estimate is eight times larger than ours.[3] If Zambia is added to both totals, the irrigation potential for all six countries is about 4.2 million hectares, which is slightly more than half the 7.9 million hectares estimated by the FAO (Table 2.3). If Zambia were to be the norm rather than the exception, and the

---

[2] This assumption would seem to be supported by data from Nigeria. In 1986 the government produced a draft master water plan, under an FAO Technical Assistance project. This plan estimates the irrigation potential of the country to be about 1.2 million hectares. This figure pertains almost exclusively to modern irrigation. Some 800,000 hectares of the 850,000 currently irrigated in Nigeria are under small-scale irrigation. Modern and small-scale irrigation are competitive in Nigeria, and the development of the former reduces the amount of water available to the latter. Therefore, if the FAO estimate of 2 million hectares (table 5.2) includes small-scale irrigation and the government estimate does not, the "true" irrigation potential of Nigeria should be somewhere between the two. The midpoint estimate, 1.6 million hectares, is 20 percent lower than the FAO estimate.

[3] Of course, there is room for dispute on such points because of a still-developing data base.

Table 2.3

IRRIGATION POTENTIAL OF SIX SUB-SAHARAN COUNTRIES
(10^3 ha)

	As estimated by FAO a/	As estimated by the study	Study/FAO's (%)
Mali	340	496	–
Sudan	3,300	2,446	–
Botswana	100	57	–
Zambia	3,500	423	12
Zimbabwe	280	460 a/	–
Kenya	350	245	–
Total	7,870	4,127	52
Total w/o Zambia	4,370	3,704	85

a/ FAO Irrigation in Africa- South of the Sahara, Investment Center Technical Paper No. 5 (Rome, 1986) Table 3.
b/ Average of potential for "commercial" and for "communal" types of development.

Source: J. Olivares-"Options and Investment Priorities in Irrigation Development", UNDP INT/82/001, Final Report, World Bank, Agriculture and Rural Development Department, August 1987, table 26, p.99.

52 percent ratio were to be applied to the FAO estimate of 33.6 million hectares, the irrigation potential of SSA might be as low as 17.5 million hectares.

As mentioned earlier, the eight countries with high irrigation potential receive substantial rainfall. If just half of their irrigation potential were to be climatically unnecessary, FAO's estimate would drop to 22 million hectares. Then, if the 85 percent proportion estimated above were to hold in this reduced figure, the overall irrigation potential for Africa south of the Sahara would be about 18.7 million hectares. In short, I would surmise that SSA irrigation potential is not much more than 20 million hectares, of which 5 million are already developed. That would leave 15 million hectares as the total area suitable for future irrigation development.

These figures are quite small, both in absolute terms and in relation to the population of the region. To fully appreciate their order of magnitude, one should compare them with absolute figures for other regions and countries. Some 200 million hectares are currently being irrigated worldwide. Large countries like India and China irrigate 40 million and 45 million hectares, respectively. The irrigation potential of a country like Brazil has been estimated to be anywhere between 24 and 48 million hectares. Therefore, the total irrigation potential for the SSA region may be barely half of the area currently irrigated in either India or China (there is no available estimate of the irrigation potential of either country) or of Brazil's potential.

The African figures are also very small when related to population. There are 380 million Africans for 20 million hectares of irrigation potential (i.e., 5.3 irrigable hectares per 100 Africans), while there are between 20 and 40 hectares per 100 Brazilians and some 17 hectares per 100 Thais. No estimate for the irrigation potential of India has been reported, but if India were today irrigating, say, two-thirds of its (undetermined) potential (the figure for Tamil Nadu is 100 percent, while that for Orissa, one of the "countries" in the World Bank Study, is 35 percent), India's irrigation potential would be some 60 million hectares. This would yield a ratio of about 7.7 irrigable hectares per Indian, or almost 50 percent more than in SSA.

By these calculations, the irrigation potential of SSA is small, and its development is not likely to make a major contribution to agricultural development on the continent as a whole. But Africa is widely heterogeneous. Although the eight countries mentioned above have plenty of rainfall and thus a large irrigation potential and little need for it, another set of (coincidentally) eight countries (Mauritania, Senegal, Mali, Burkina Faso, Niger, Somalia, Kenya, and Botswana) face the opposite situation. They have little or no land with a significant rainfall, rather low irrigation potential (the figures for Mali, Kenya, and Botswana are 500,000, 245,000, and 57,000 hectares, respectively), and are no longer able to sustain their current population with low-input rainfed farming. For these countries--which contain 14 percent of the population of the region--irrigation is likely to be an essential element of any national development policy. But since they lack rainfall, water resources available for irrigation are limited. FAO estimates their total irrigation potential combined at a bare 1.5 million hectares, so that their development, however important locally, would not have much effect on the overall region. Similarly, even the most humid countries in SSA have some arid areas in which irrigation development will be crucial, but again it will only be important at a local level.

The Cost Of Developing Irrigation In SSA

The financial dimension. Developing SSA's irrigation potential has always been considered a high-cost proposition. Figures like US$40,000 per hectare, which is the current estimate for one project in Kenya, are often quoted in this regard. It is true that developing irrigation in Africa has proved expensive, but there are some good reasons for that (see the annex to this chapter). In particular, "irrigation" costs usually include many elements that should not really be listed as such. The most obvious examples are roads, houses, electric grid, and public service facilities. Many irrigation projects in Africa have had to include such investments because of the extremely low population density--or absolute absence thereof--in the areas to be developed. They represent nation- or region-building costs, which should not really come under the "irrigation development" label.

Irrigation costs for seven countries in the region, plus for two others, are still quite large in comparison with other regions. According to FAO, large-scale, government-developed project costs have run between US$5,000 and $10,000 per hectare for Mali, Burkina Faso, Madagascar, and Mauritania; and between US$10,000 and $15,000 per hectare, and even up to $20,000 per hectare for Ghana, Niger, and Nigeria. In Kenya, the average for major new, fully controlled projects is $8,000 per hectare (in 1985 U.S. dollars). (Unfortunately, FAO does not specify whether these cost data include only the "pure" irrigation components or cover "complete" project costs, including nonirrigation components.) In contrast, small-scale schemes would cost between US$1,500 and $2,500 per hectare in Ethiopia, and less than US$1,000 per hectare in Mali.

The World Bank study provides the estimated cost of developing every possible project in a country. Expressed in 1984 U.S. dollars, investment costs in Sudan range from $180 to $5,800 per hectare, with a weighted average of $2,850 per hectare (cost per project weighted by the respective project area). Data on project costs in Mali do not include the cost of the dams--which may be a substantial share of total project costs--either because they were already built and the expenditure had to be treated as a sunk cost, or because no data were available on the additional dams required. The cost of the remaining components ranges from $240 to $7,900 per hectare, with a weighted average of $1,560 per hectare. Weighted averages for Botswana and Kenya are $5,900 and $5.600 per hectare, respectively; $2,000 per hectare for Zambia; and $9,500 per hectare for Zimbabwe (all in 1985 U.S. dollars). On average, these figures are slightly lower than the US$8,000 quoted by the FAO.

One analysis currently in progress is examining irrigation projects throughout the world that were supported by the World Bank between July 1973 and June 1985 (49 new projects and 76 rehabilitation or modernization projects).[4] Preliminary results indicate that once project size, pluviometry, national per capita income, and year of Board approval are taken into account, the cost of the irrigation projects supported by the World Bank in SSA was not significantly higher than in other regions.

The environmental dimension. The cost of developing irrigation in SSA cannot be measured in financial terms only. More than in any other region, irrigation development in SSA has proved to have substantial environmental costs. As in any other area, it may lead to water-logging, salinization, damage to fisheries or water supply systems downstream, or other similar problems. Two repercussions that have been particularly severe in SSA are (1) the associated development of waterborne and water-related diseases, and (2) competition with floodplains.

Waterborne and water-related human disease exists in all areas where irrigation is practiced. In Africa, however, for reasons not fully understood, the scale of disease connected with irrigation is massive. The most serious threat is schistosomiasis (bilharzia). Horror stories are told in almost every country. The rates of prevalence among the population at risk (who, ironically, are often called "beneficiaries") went from less than 5 or 10 percent before, to more than 80 percent after irrigation in some extreme cases like Lake Volta or the Gezira. The World Bank has been financing the required health projects in some cases. Malaria is a distant second, but still a significant threat. Fortunately, other serious diseases like Japanese encephalitis have not yet made their appearance in Africa. On the positive side, the incidence of onchocerciasis (river blindness, a disease the World Bank has also been helping to control) is reduced with the expansion of irrigation.

If human suffering cannot be quantified, an effort should at least be made to quantify the value of the healthy days of life lost or the number of working days lost due to the increased incidence of the diseases associated with irrigation. These amounts should then be added to project costs. A methodology for assessing ex-ante the human health risks associated with irrigation development--which would allow engineers to design projects so as to minimize or prevent such risks--was developed and tested under the World Bank study.

The floodplains problem arises because large shallow plains exist in many river basins in Africa. They accommodate seasonal floods, and therefore greatly expand during the flood season and thereafter shrink in size and in depth. Some 25 of them are of significant size (larger than 100,000 hectares each; altogether they cover some 30 million hectares). The best-known among the in-stream ones are the Sudd, the largest swamp in the world (9 million hectares), on the White Nile, and the Inner Delta of the Niger. Among the terminal ones are

---

[4] J. Olivares and J. Hammer, "The Cost of Irrigation Development," in preparation.

Lake Chad on the Longone-Chari system, and the Okavango Delta (Table 2.4).[5] Floodplains are associated with many benefits: hydrological (flood control and flow stabilization, sediment trap), economic (fish, beef, and other meat production; rice; other grains), social (large populations live and earn their living there), and ecological (large populations of wildlife, particularly migratory birds, live or sojourn there; they are also important reservoirs of vegetal diversity), to mention just a few. They "consume" (basically by evaporation) huge amounts of water (estimated to be two-thirds of the incoming flow in the case of the Sudd, and half of it in the Niger Inner Delta), which could otherwise be used to irrigate land, generate power, or supply water to populations. They "transform" this water into substantial benefits (some 1.5 to 2 million cattle live in these areas today; 100,000 tons of fish are caught in the Inner Delta of the Niger every year).

Table 2.4
FLOODPLAINS LARGER THAN 1,000 km² IN SUB-SAHARAN AFRICA

Drainage System	Name	Country	Area (km2)	Wetland Type	Special Features	Chapter of the present report
Zaire/Congo	1. Middle Congo Swamps	Zaire, Congo	40,550	Riverine swamps and floodplain	Permanent and seasonal	4
	2. Mweru (Luapula River)	Zambia, Zaire	4,580	Shallow lake and swamps	3-70 m deep.	5
	3. Bangweulu	Zambia	8,800	Shallow lake, swamps, and floodplain	Lake, 4m deep	6
	4. Kamulondo Depression	Zaire	11,800	Shallow lakes, swamps, and floodplains Lualabe River.	Lake Upemba (550 km2) and 50 other lakes on floodplain of	7
	5. Malagarasi	Tanzania	7,357	Floodplain and swamps		8
Niger-Benue System	6. Niger Central Delta	Mali	20,000	Floodplain, swamp, and shallow lakes	Lake Niangaye is 402 km2; rice is grown.	9
	7. Benue River	Cameroon/Nigeria	3,100	Floodplain		10
Nile System	8. Machar Swamps	Sudan	6,700	Swamps and floodplains		11
	9. Sudd	Sudan	92,000	Swamps and floodplains	Sudd is largest single swamp area in the world.	12
	10. Kenamuke Swamp	Sudan	12,995	Swamp		13
	11. Lotagipi Swamp	Sudan, Kenya	12,900	Swamp	Internal drainage basin.	13
	12. Lake Kyoga	Uganda	2,700	Shallow lake and swamps	Average depth 6 m	14
Zambezi System	13. Marotse Plain	Zambia	9,000	Floodplain	Heavily cultivated and grazed	15
	14. Liowa Plain	Zambia	3,500	Floodplain and swamp	Important wildlife area.	16
	15. Lukanga Swamp	Zambia	2,500	Swamp.		17
	16. Kafue flats	Zambia	6,000	Seasonal swamp, floodplain	Flood regime threatened by Itezhitezhi Dam.	18
Western System	17. Volta River	Ghana	8,532	Floodplain		19
	18. Senegal River	Senegal, Mauritania	4,560	Floodplain	Important fishery.	20
Eastern System	19. Kilombero Floodplain	Tanzania	6,647	Floodplain and swamp.	Highly productive fishery	21
	20. Rufiji Floodplain	Tanzania	1,450	Floodplain	Active fishing; hydrology to be changed by Stieglers Gorge Dam	22
	21. Shire Floodplain	Malawi	1,030	Floodplain and swamp		23
Chad System	22. Lake Chad	Chad, Niger, Nigeria	13,800	Shallow lake and swamps	Rich fishery.	24
	23. Yaeres	Chad, Cameroon	5,957	Floodplain	On Chari and Logone Rivers.	25
Okavango System	24. Okavango Delta	Botswana	16,000	Swamp		26

Source: A.H. Pieterse, et al, "A Resource Planning Data Review on Major African Inland Swamp and Floodplain Ecosystems", Royal Tropical Institute, March 1987, Table 1.

---

[5] A. J. Pieterse et al., _A Resource Planning Data Review on Major African Inland Swamp and Flood Plain Ecosystems_ (Amsterdam: Royal Tropical Institute, March 1987).

Irrigation development in these basins may reduce water inflows into the swamps (e.g., irrigation in the Niger Upper Valley, the Hadejia-Jama'are Basin in Northern Nigeria, or the Chari River in Northern Cameroon). In some cases it may be necessary to bypass them in order to reduce water evaporation (e.g., in the Jonglei canal on the Sudd), or to extract water from the directly (e.g., in the South Chad project in Northeast Nigeria and polder development in Chad). With the ensuing reduction in the swamp area, however, the level of benefits they currently generate would drop, wildlife would be jeopardized or destroyed (e.g., the Kafue flats are the only known habitat for the Kafue lechwe; the swamps on the Senegal and Niger rivers are crucial resting areas for the migratory birds of Western Europe); the livelihood of large numbers of people would be seriously affected (some 2 million persons live in and around the African floodplains); and social and international conflicts could be expected to develop.

No one has yet attempted to estimate the overall worth of the benefits forgone in association with irrigation development in basins where floodplains exist. However, water diverted from swamps may find worthwhile uses elsewhere, as illustrated by a comparison of the productivity per unit of water in the Inner Delta of the Niger and the neighboring Office du Niger irrigated rice scheme. A recent report indicates that the former is producing 10,000 tons of meat, 120,000 tons of milk, 100,000 tons of fish, and 80,000 tons of rice per year, whereas the latter produces 100,000 tons of rice. When computed per unit of water, the irrigated rice produces almost 10 times as many calories and almost twice as much protein as the swamp, although the swamp's produce is more varied in type and nutritional value.[6]

Conclusion

The irrigation potential of Sub-Saharan Africa may be no greater than about 20 million hectares, 5 million of which are already developed. Therefore irrigation is unlikely to play a significant role in the development of the overall region. It may be important in individual countries, however, particularly those on the fringes of the Sahel.

Contrary to popular belief, it would not necessarily be more expensive to develop Africa's irrigation potential than is the case in other regions, once the relevant variables (project size, pluviometry, per capita income) are taken into account. However, the development of irrigation in SSA might incur high environmental costs, particularly in the form of increased water-related diseases and the threat to the vast (and often very productive) floodplain ecosystems. The methodology developed by the World Bank for assessing ex-ante the human health risks associated with irrigation development should help engineers minimize or prevent such risks in their project designs.

---

[6] C. A. Drijver and M. Marchand, Taming the Floods, Environmental Aspects of Floodplain Development in Africa, (Center for Environmental Studies, State University of Leiden, 1985).

ANNEX

The following extract is taken from FAO, <u>Irrigation in Africa South of the Sahara</u>, Investment Centre Technical Paper 5 (Rome, 1986), Annex 1, pp. 146-48.

<u>External</u> causes of high investment costs include:

- The overvaluation of most African currencies, which inflates all costs in dollar terms.

- Difficult access to the landlocked African countries and to most irrigation perimeters. In Mali, for instance, cement costs 40 to 50% more at Gao than at Bamako. Poor roads in the Zone Lacustre add 15%, but the cost would be 25% higher than at Bamako due to the transport distance alone.

- Taxes, which unnecessarily raise certain costs--for instance a 15% wages tax in Mali, plus import duties and fuel taxes--and which are rarely waived.

- The lack of local manufacture of equipment and spares together with supply difficulties, make it necessary for projects to carry heavy stocks; when manufacturers make technical modifications considerable stocks of spares become obsolete, increasing operating costs.

- The lack of local equipment sales and service agents. For instance to obtain a pump adapted to the conditions of Mali, with an efficiency of 70% as against the 40% of locally sold pumps, it is necessary to purchase in Europe.

- Shortage of skilled local personnel (mechanics, construction workers) and of small contractors, especially for earthworks.

- The use of tied external funds to build irrigation administration costs and, especially, the purchase of nonstandard equipment requiring special maintenance and spare parts.

There are other <u>physical</u> causes in addition to distance.

- Major flood protection dykes are necessary for most rice perimeters. In many Asian countries, such dikes were built long ago and no longer appear as investment costs, whereas they often account for one-quarter of the cost of civil works in West Africa.

- The patchy distribution of irrigable soils and uneven shape and topography of many African perimeters, which calls for complex water distribution and drainage networks with considerable leveling.

- The low population density, which sometimes requires associated investments for colonization and other social infrastructure.

- The reservoirs and dams which are essential to stabilise the erratic flows of West African rivers. Because of the predominantly flat local topography dams are usually long, low and expensive, reservoirs are shallow, and evaporation is high. Hence a large investment is required to store a usable volume of irrigation water.

♦ There are no abundant, shallow sources of groundwater suitable for localised irrigation or conjunctive use comparable to those of India or Pakistan. Maximum output of boreholes often is less than $5 m^3/hr$ and water often has to be pumped from 10-m depths or more.

♦ It is often necessary to include related investments for land clearing and access tracks, which increase total scheme costs compared with other countries.

♦ The climate is severe. The possibility of very intense rainfall and cyclic droughts require high safety coefficients in project design.

The external causes of high costs, if they are to be reduced at all, would mostly require government policy changes or will only come about as a product of overall industrialisation and further national economic development. Most of the physical causes are beyond human intervention altogether. However, several of the internal causes of high costs could be reduced by the irrigation agencies themselves. Principal internal causes are as follows.

♦ Most studies are made by foreign consulting firms which are not subjected to adequate local control. This results in three kinds of extra cost. First, such studies usually cost about twice what they would if they could be made by local firms; hence they add an extra 2 to 4% to investment costs. Second, the consulting firm's reputation is at stake, so that often superfluous design precautions are taken to make sure that the structures concerned will last a long time. Exaggerated safety measures in civil engineering involve considerable extra costs and few of these in the irrigation agencies who order a job are technically competent to propose simpler standards to specialised foreign consulting firms. Third, foreign firms may have little local experience and do not always design structures in line with local resources. In Mali, for instance, 3 m wide hand-operated control gates with wooden frames can be manufactured locally. If 5-m wide gates are specified that have to be made in Europe, as well as the frames which must then be of metal, and they need a gantry to operate them. As control structures typically account for 20 to 25% of project cost a 20% saving on items such as these saves 4 to 5% of the total investment.

♦ Simplified initial designs could be adopted, even if they may need repair or upgrading after a year's operation. For example, it was noted in Mali that some parts of dykes had been weakened by wave action. However, rather than designing all dykes with full protection it was found preferable to build simple dykes and to protect in the following year only the parts that had been weakened. This method obviously requires prior agreement with the funding source that money will be kept available for subsequent upgrading, and that no one should criticise the builder of the dyke when a part is damaged in the first year.

♦ If local consulting firms are enlisted, the funding sources must, however, have confidence in them, help them with consultants' mission, and do away with certain useless controls.

- Common design standards could be agreed, and departure from these standards should have to be specifically justified by project planners. In this way costly over-designing could be reduced. Among the sources for this study, for instance, the following were noted for projects designed for comparable ecological conditions:

    - Strickler coefficients varied from 25 to 40, sometimes for the same agency, resulting in very different channel sizes for a given duty.
    - Irrigation duties for flood-irrigated rice ranged from 1.54 to 10 liters per second per hectare.
    - Drainage duties ranged from 0.69 to 17 liters per second per hectare.

- Irrigation agencies could do more to standardise their equipment. To do so it would be necessary to overcome obstacles due to tied aid, plus the not infrequent local problems posed by agents or suppliers with special influence.

- Project designers should be required to draw up detailed manuals for irrigation scheme operation and maintenance at the design stage. This would focus the designers' thinking on maintenance needs, and perhaps cause them to adapt the project accordingly.

- There is no communication among countries on technical improvements for cutting costs. The following are examples from Mali which might with advantage be applied elsewhere in the region:

    - laterite coating of dykes, even though normally left uncompacted, provides an all-weather track and reduces maintenance cost;
    - very cheap types of prefabricated concrete channels have been developed;
    - masonry has been used for some structures instead of reinforced concrete, resulting in considerable savings;
    - drainage ditches are sown with bourgou, a plant which is regularly cut for fodder by the local population, thus ensuring free drain maintenance.

- Owing to currency overvaluation and low population density labour can be expensive in Africa. Nevertheless, to reduce dependence on imported equipment and fuel it may still be advisable to adopt more labour-intensive construction methods, as is done in India. At present such an approach is usually limited by a shortage of skilled supervisors and by disincentive labour regulations.

- Lastly, government costs could be lowered by greater participation of beneficiaries in construction and O & M. Improved security of tenure of irrigated land, under a revised legal framework, could be a major incentive towards improving farmers' participation.

CHAPTER 3: PRIVATE SECTOR IRRIGATION IN AFRICA

David Seckler
(Director, Agricultural Policy and Resource Development,
Winrock International Institute for Agricultural Development)

In writing this chapter, I became aware of four aspects of irrigation that I had not adequately appreciated before. First, private sector irrigation "focuses the mind wonderfully," as Dr. Johnson said of the prospect of being hanged, on the private, financial return to investment in irrigation. Although a favorable return is not a sufficient condition for private sector investment in irrigation, it is, to say the least, a necessary one. The ultimate criterion for investing in public sector irrigation is the social costs and benefits. Although this is also the case for private sector irrigation, the additional challenge there is to get the private sector to reflect social costs and benefits by actually investing in socially desirable activities, such as irrigation. This is the true and original meaning of "getting prices right," and it is not an easy thing to do.

Second, I found that the return to investment in irrigation is largely determined by the physical, social, and economic <u>environment of irrigation</u>. I now understand that Asia had a favorable irrigation environment when I was working there, and that I, like other Asian irrigation <u>wallahs,</u> could concentrate on the technological, institutional, and economic details of irrigation. If all this were properly worked out, the environment of irrigation would take care of the rest. In much of Africa, however, we are working in an unfavorable environment. We first have to understand that environment and find out how to change it before we can be confident that private sector irrigation, or any other kind of irrigation, will succeed there.

At least six environmental conditions must be satisfied if private sector irrigation is to be feasible: (1) the conversion of rainfed agriculture to irrigated agriculture must yield sufficiently large increases in agricultural production to justify the additional costs of irrigation; (2) there must be a supply of good-quality water and irrigable land within reasonable proximity to each other; (3) irrigation requires a reliable and economical supply of other inputs such as improved seeds, fertilizers, and labor to realize its productive potential; (4) there must be an economically accessible market for the purchase of these and related inputs and for the sale of the marketable surplus from irrigated areas: (5) there must be a reasonable transportation system between the irrigated area and market centers; and (6) producer input/output prices must be "right" and reasonably stable over the life of the irrigation investment. Even someone only casually acquainted with the irrigation environment of Africa cannot fail to see that these conditions are not adequately satisfied in many places in Sub-Saharan Africa (SSA).

The importance of the irrigation environment is perhaps best illustrated in what is rightly considered the most successful case of private sector irrigation in developing countries: private sector tubewell irrigation in India, Pakistan, and Bangladesh. Private tubewells have flourished in areas with reasonable roads, research and extension systems, crop price-support programs, credit and (usually) subsidies for irrigation equipment, fertilizers, and energy. Also, private tubewells have largely developed in and around the command areas of large surface irrigation systems. There are three reasons for this: (I)I deep percolation losses from the surface systems recharge the aquifers for tubewells; (2) the tubewells are often used together with surface irrigation water, which lowers pumping costs and concentrates these costs in periods of highest marginal returns; and (3) the tubewells ride piggyback on the infrastructure created for the surface systems. I believe that most of the problems of irrigation in Africa are not problems of irrigation per se, but problems of the environment of irrigation--hence the considerable amount of attention given to this problem below.

The third point to note about irrigation is that caution must be exercised in using the phrase "private sector irrigation." If we apply it only to individually owned and operated systems, we unduly restrict the range of technological and institutional options available to us. Given the generally small size of farms in Africa, only pump irrigation technology is sufficiently divisible to be wholly owned and operated by individual farmers. This technology is restricted to land over aquifers and along the banks of rivers and lakes. Although this kind of small-scale, individually owned and operated system is nearly ideal, it has only limited scope for realizing Africa's irrigation potential.

Other irrigation technologies are "public goods" in the sense of joint production functions; that is, the same resource is used by more than one individual. It has therefore been natural for public agencies to own and manage irrigation systems. I am not a member of the chorus that sings damnation of all public irrigation systems. Some (for example, the vast warabandi system of northwest India) perform exceptionally well, but many perform very poorly. Between the purely public and purely private systems lie a variety of nongovernmental, intermediary institutions, both of a commercial and philanthropic nature, that hold great promise for the development of private sector irrigation in Africa.

Fourth, the generally unfavorable irrigation environment in Africa leads me to believe that a "growth centers" strategy would be the most effective means of developing irrigation in this region. Specific areas of Africa could be selected as irrigation-based growth centers, beginning with those that have high irrigation potential and a reasonable infrastructure of roads, markets, and input supplies already available. Once these first-order growth centers are well on the way to development and the lessons on how to develop them have been learned, new second-order centers could be established about 100 kilometers away in more backward areas. The first- and second-order growth centers would be connected by transportation and communication facilities. Thus, the growth centers could develop symbiotic trade relations that would stimulate growth in the areas between the centers. Then third-order growth centers would be established farther out in the hinterland, and so on.

This irrigation-based, growth-center development process is highly suited to the agroclimatic conditions in Africa. Indeed, development could proceed right across the northern part of Sub-Saharan Africa, from Nigeria in the west and Kenya in the east, to create an integrated, coast-to-coast, "mid-African growth corridor." Areas in the northern part of the corridor would be primarily irrigated areas, while those to the south would develop mainly through fertilizer and other inputs to rainfed agriculture. Eventually the growth corridor would be supported by a transcontinental transportation system, including a railway, highway, natural gas lines, and communication facilities. With duty-free transit privileges across countries and reduced tariffs, this area would develop rapidly, both agriculturally and industrially. Similar growth corridors could be established in other high-potential areas of Africa.

The Synergism Between Irrigation And Other Factors Of Production

Irrigation is not a "stand-alone" technology. Although it helps to increase agricultural production by reducing water stress on crops, high returns depend on other factors as well, such as adequate use of fertilizers, good germplasm, and labor. If these inputs are available, and other conditions such as prices are favorable, the complete package of inputs needed to realize the high returns will be present, and investors will be more likely to consider irrigation ventures. In other words, irrigation promotes the use of other inputs, and other inputs promote the use of irrigation in a synergistic process of agricultural development.

Figure 3.1 illustrates how the process works in some countries of Asia. In the left quadrant of Figure 3.1, fertilizer consumption is shown in relation to a "water regime index," which, as explained below, is an index of the soil moisture regime based on precipitation and irrigation. The right quadrant shows the relationship between fertilizer consumption and cereal yield. (The upward-sloping curve reflects physical productivity of fertilizer.) In large part, variations in the water regime account for variations in fertilizer use, and variations in fertilizer use account for variations in yield among the countries.

This example demonstrates two fundamental facts about agricultural production that are sometimes overlooked: high yields cannot be obtained without ample fertilizer and good germplasm; and the high use of fertilizer is not economically viable in poor and unstable water regimes. These facts help to explain the differences in agricultural production in Asia and Africa. In Asia, more than 50 percent of the arable land is irrigated, in contrast to less than 5 percent in Africa (half of which is in northern Africa). Also, Asia consumes about 40 kilograms of fertilizer per hectare of arable land per year, whereas Africa consumes less than 10 kilograms.

It cannot be that Africa needs less irrigation or fertilizer than Asia does, nor that Africa has a surplus of agricultural land and therefore requires a low-input, low-yield form of production. As the FAO (1986, p. 13) rightly observes, the idea that Africa has large land and labor surpluses is a myth. Rather, the problem of agricultural development in Africa seems to be how to get irrigation and fertilizer moving. The FAO report cited above suggests that <u>neither</u> the green revolution technology used in Asia nor irrigation are the ultimate answer, as their potential effectiveness in Africa constitutes another myth.

Two reasons are given for the "Asian technology myth." First, the green revolution works "only when irrigation or reliable rainfall provides sufficient moisture." Second, the rainfed, high-yielding varieties (HYVs) so successful in Asia (sorghum and millet) have been difficult to cultivate in Africa. Local varieties have higher yields under rainfed conditions and have superior cooking qualities. It is also said that the irrigation myth has grown up because most of Africa's water resources are in areas where irrigation is not needed. Africa is said to have sufficient water to irrigate 40 million hectares, but almost half of this is in areas with ample rainfall. Thus the feasible potential is about 20-25 million hectares, and this only in the long term. Moreover, irrigation may not be economically viable in Africa because irrigation with full water control tends to cost considerably more than in India.

Indeed, the Asian technology myth erroneously associates the success of agricultural production in Asia with HYVs. These varieties were important in the later stages of the development process, but the first step was to test irrigation and fertilizer consumption with <u>local varieties</u>. It was not until the production possibilities of the local varieties were <u>exhausted</u> that the HYVs came to play a vital role and once the base of irrigation and fertilizer use <u>had already been established</u>.

Even if it is true that only 20 to 25 million hectares are suitable for irrigation, this still amounts to a great deal of irrigation. It is equal to about half the irrigated area of India, which has nearly twice the population of SSA, or about the same irrigated area per person. With the increased stability of food production under irrigated conditions and the large

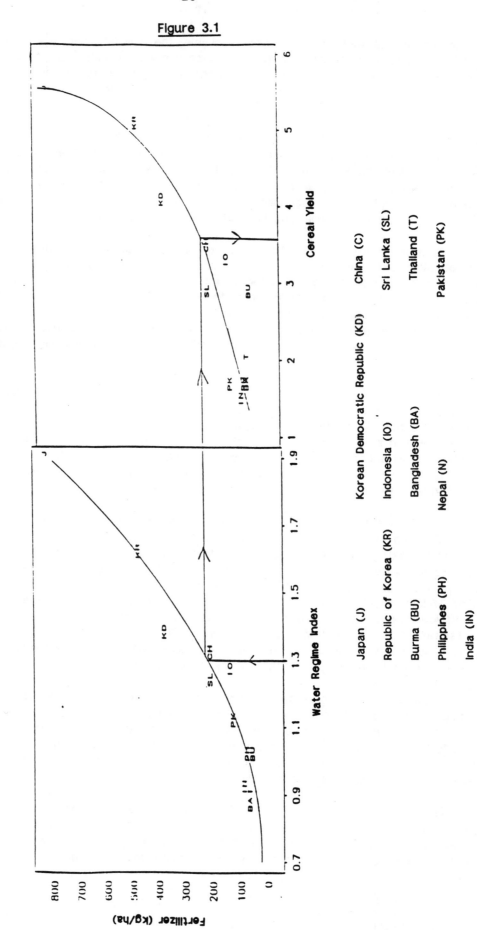

Figure 3.1

potential for rainfed agriculture, the full development of Africa's irrigation potential could provide food security for the entire continent. Also, there is no inherent reason why irrigation in Africa should cost more than it does in Asia. With the appropriate choice of irrigation technologies, as discussed below, and the attention to mistakes of Asia in this regard, it could cost less.

These issues have considerable bearing on the prospects for private sector irrigation in Africa. If areas with irrigable land and sufficient water do not need irrigation, and areas that do need irrigation do not have these resources, there is little scope for irrigation of any kind. Also, it is vitally important in developing private sector irrigation to know precisely which areas of Africa promise a reasonable return to investment in irrigation. The next two sections examine these agroclimatic issues.

Irrigation Needs And Potential in Africa

Map 3.1 gives an idea of the physical scope of our discussion, while maps 3.2 and 3.3[7] provide a broad picture of irrigation needs and potential in Africa. Map 3.2 shows the agroclimatic zone that is suitable for maize and other cereal crops under rainfed conditions. The blank areas show where crops cannot grow without irrigation. Note the rather large area in which irrigation is not required under rainfed conditions. Note, too, that nearly half of this area lies in the central rainforest, most of which is considered unsuitable for crops in any case. Thus, contrary to the irrigation myth, most of the area that is suitable for rainfed crops in Africa <u>requires irrigation</u>.

Map 3.3 shows the areas in which there is sufficient water to work up irrigable lands. This map is encouraging: much of the area that needs irrigation can be irrigated. Actual water availability may, in fact, be substantially underestimated, as recently discovered in the irrigation schemes operating in Asia. Suppose that only 50 percent of the available water is used for irrigation and that the rest is lost to surface drainage or deep percolation (Map 3.3). In fact, this water is lost only in the first iteration of the irrigation system. In following iterations, the water can be picked up from downstream subsurface flows, or pumped up from aquifers, and reused for additional irrigation. That is to say, although each <u>part</u> of the irrigation system may be performing at low efficiency, the system <u>as a whole</u> may be highly efficient. If water recycling is taken into account, irrigation potential may be considerably higher than Map 3.3 suggests.

One problem that arises here is that areas "suitable," or even "very suitable," for maize and other crops under rainfed conditions may <u>not</u> always be the areas where irrigation is not required (see Map 3.2). Apparently although some areas do not require irrigation (in the sense that production response would be negligible), other areas that are suitable for rainfed crops would also benefit substantially from irrigation. Consequently, the suitability or unsuitability of areas cannot be meaningfully defined in absolute terms. Rather, it must be defined in terms of a continuum of greater or lesser productivity of specific crops under varying conditions of water availability. Since the return to investment in irrigation depends largely on the difference in agricultural production under rainfed and irrigated conditions, these differences must be defined rather closely, as illustrated in the next section.

---

[7] Food and Agriculture Organization, <u>Irrigation and Water Resources Potential for Africa</u> (Rome, 1987).

Water Deficiency Factors

Buringh, van Heemst, and Staring [8] have found that soil and water deficiency factors can be used to roughly evaluate whether agroclimatic areas are suitable for rainfed or irrigated agriculture.[9] Briefly, the water deficiency factor (WDF) is an index of the probability that annual crop production will not reach its economic optimum in a given agroclimatic zone owing to water constraints. If WDF = 1.00, no loss is expected due to water constraints. If WDF = 0.50, annual crop production could be economically doubled with perfect irrigation, and so on. Buringh et al. have also estimated a soil deficiency factor (SDF) by the same process. Since one of the major functions of soil is to hold water, this constraint is generally not so severe for irrigated as for rainfed areas.

The WDF of an agroclimatic zone can be used as a basis for estimating whether it pays to convert from rainfed to irrigated agriculture. Broadly speaking, for example, private sector pump irrigation must yield an expected net additional return per hectare of crop of around 30 percent before it is feasible. This implies a gross return, or additional yield, of about 40 percent. Thus, irrigation is likely to be feasible where, at the upper bound, WDF is less than 0.60.

There is also a lower bound for WRI under most conditions. With the exception of large river or storage/canal systems, local irrigation projects depend mainly on local precipitation for water supply, and so the area should not be too dry. Also, pump irrigation costs vary with the amount pumped, so it is better to supply only a portion of crop water requirements by pump irrigation in the periods when irrigation is most productive. Thus, areas with a WDF below 0.30 are likely to be to dry for local systems. These guidelines can be followed to get an approximate idea of the areas most suitable for irrigation development in Africa.

Map 3.4 shows agroclimatic zones and accompanying deficiency factors from the study by Buringh et al, and Map 3.5 provides similar details for Asia. It is interesting to note that zone B4 occurs both in the northern part of sub-Saharan Africa and in the highly productive area of northern India. Map 3.5 shows the areas in Africa where the WDF is between 0.61 and 0.29, and the soil deficiency factor is also favorable. These are judged to be the most suitable areas on the basis of productivity of irrigation. Map 3.3 indicates that in most of these areas, water is available for more than 50 percent of the irrigable land. Although these are rough estimates of large-scale agroclimatic zones, they do give a good idea of the areas most suitable for irrigation development in Africa.

Table 3.1 shows irrigated areas and the potential for irrigation by countries in SSA. The overall potential appears to be upward of 33 million hectares. Some of this potential exists in areas where irrigation is not needed or feasible, (such as parts of Zaire), and so this table does not necessarily contradict the FAO figures cited earlier. Note that the largest potential for irrigation exists in Angola.

---

[8] P. Buringh, H. D. J. van Heemst, and G. J. Staring, Computation of the Absolute Maximum Food Production of the World (Netherlands: Agricultural University, Wagingen, 1975).

[9] David Seckler and John LaBore, "Agroclimatic Indices for Asia and Africa," Report to the Development Studies Center, Winrock International Institute for Agricultural Development (Arkansas, 1988).

Table 3.1: SUB-SAHARAN AFRICA: ESTIMATES OF IRRIGATED AREAS, 1987
IN RELATION TO IRRIGATION POTENTIAL

Country	Irrigation Potential ('000 ha)	Area Developed 1982 (thousands of ha)			Developed as % of Potential
		Modern	Small-Scale or Traditional	Total	
Angola	6,700	0	10	10	< 1
Benin	86	7	15	22	26
Botswana	100	0	12	12	12
Burkina Faso	350	9	20	29	8
Burundi	52	2	50	52	100
Cameroon	240	11	9	20	8
Central African Republic	1,900	0	4	4	< 1
Chad	1,200	10	40	50	4
Congo	340	3	5	8	2
Equatorial Guinea	n.a.	n.a.	n.a.	n.a.	n.a.
Ethiopia	670	82	5	87	13
Gabon	440	0	1	1	< 1
Gambia	72	6	20	26	36
Ghana	120	5	5	10	8
Guinea	150	15	30	45	30
Guinea Bissau	70	n.a.	n.a.	n.a.	n.a.
Ivory Coast	130	42	10	52	40
Kenya	350	21	28	49	14
Lesotho	8	0	1	1	13
Liberia	n.a.	3	16	19	n.a.
Madagascar	1,200	160	800	960	80
Malawi	290	16	4	20	7
Mali	340	100	60	160	47
Mauritania	39	3	20	23	59
Mauritius	n.a.	9	5	14	n.a.
Mozambique	2,400	66	4	70	3
Niger	100	10	20	30	30
Nigeria	2,000	50	800	850	43
Rwanda	44	0	15	15	34
Senegal	180	30	70	100	56
Sierra Leone	100	5	50	55	55
Somalia	87	40	40	80	92
Sudan	3,300	1,700	50	1,750	53
Swaziland	7	55	5	60	>100
Tanzania	2,300	25	115	140	6
Togo	86	3	10	13	15
Uganda	410	9	3	12	3
Zaire	4,000	4	20	24	1
Zambia	3,500	10	6	16	< 1
Zimbabwe	280	127	3	130	46
Total	33,641	2,638	2,381	5,019	14.9

Sources: Study team estimates of areas developed; irrigation potentials from FAO Land and Water Division, 1985 (provisional estimates).

Irrigation Cultures, Technologies, And Institutions

There is a large menu of irrigation technologies from which to choose, each appropriate for different agroecological conditions, and each requiring a particular institutional structure, which may or may not be compatible with local cultures. The design and evaluation of irrigation systems should therefore consist of three stages. The first should be to assess the physical and cultural components of the irrigation environment, as they dictate the <u>design parameters</u> of the system. The next stage is to define the technological and institutional <u>options</u> that are possible under these parameters. The third step is to employ economic, environmental, social, and other evaluations to <u>choose</u> the optimal (the best under the circumstances) irrigation system from these options.

Some of the following discussion draws on experience with irrigation technologies and institutions in India. This is because India arguably has more experience with a greater variety of irrigation systems, especially private ones, than any other country. What has been learned in the school of hard knocks in India, both pro and con, can be useful in developing irrigation in Africa.

<u>Cultures</u>. Irrigation possesses a distinct "culture" that is not compatible with some other cultures. People from pastoral cultures, for example, have great difficulty adjusting to irrigation because they are required to change their basic life styles. The magnitude of the problem depends on the specific culture and even on differences between individuals within the same culture. One of the most gratifying, and even surprising, features of the World Bank-sponsored Semry Rice Project in Cameroon [10] is how readily local tribal people adapted, learned, and prospered under irrigated agriculture. Other cases could probably be cited where irrigation failed because of cultural constraints, although I personally cannot cite one. The cultural factor is of greater importance in the design of the management system for irrigation, which also affects the design of the physical system. Some cultures are more community oriented, whereas others are more individual oriented. Whatever the case, irrigation systems should be adapted to the local cultures, rather than the reverse--as is too often the case.

It is also vital to have professionals and citizens from the local culture involved in <u>decisionmaking</u> in all stages of irrigation design and implementation. There is no question, for example, that the Sukhomajri project in India--a highly successful scheme--would have been a failure had this principle not been religiously followed.

<u>Technologies</u>. Irrigation technologies consist of combinations and permutations of four basic elements: (1) the <u>source</u> of the water supply, (2) the <u>motive force</u> to control the water, (3) the <u>conveyance</u> system from the source to the irrigated area, and (4) the <u>application</u> system that waters the crop. These components and their subcomponents generate a variety of irrigation technologies, listed below in increasing order of cost and technical sophistication:

---

[10] World Bank, "Cameroon Second Semry Rice Project," Project Performance Audit Report (Washington, D.C., 1984).

(1) Source of water supply:

    Aquifers
    Small water-harvesting tanks

    Rivers and lakes
    Large reservoirs

(2) Motive force:

    Gravity
    Pumps
    Barges
    Animal or human power

(3) Conveyance systems:

    PCV pipes
    Canals and channels (lined and unlined)

(4) Application systems:

    Flood, furrow, etc.
    Sprinkler
    Drip

Keller [11] has identified more than 20 systems on the basis of the conveyance and application characteristics alone.

Institutions. There are also many types of institutional arrangements for financing, constructing, operating and maintaining, and using irrigation systems. These institutions depend largely on who owns land and manages the system, and how they relate to the user, usually the private farmer, and to its four basic components. Institutional arrangements may be classified as (1) individually owned and operated systems, (2) market systems, (3) water users' associations/cooperatives, or (4) government agencies. As a general policy, it is best to start with the least costly and sophisticated technological and institutional system—to stay as high up on these lists as possible, until it is necessary to go lower down. According to this rule, the most promising irrigation systems for Africa are (a) pump irrigation systems where water is available in aquifers, rivers, and lakes; and (b) water harvesting tanks in hilly areas with reasonably high but erratic levels of precipitation. These technologies also are compatible with the most effective institutional arrangements: individual ownership, water markets, and water users' associations.

Pump Irrigation Systems

As noted earlier, pump systems are the only sufficiently divisible technologies that can be owned and operated by a small farmer. However, this method cannot be used unless the

---

[11] Jack Keller, "Modern Irrigation in Developing Countries," International Commission on Irrigation and Drainage, Special Session Proceedings of the 1990 Conference (Brazil, 1989), in press.

means exists to lift water from rivers, lakes, and aquifers to the land. Some systems are human or animal powered, but these have only a small niche in irrigation. They are only appropriate where small quantities of water need to be lifted short distances to avoid temporary moisture stress. The trouble with these lift systems is that the human or animal energy expended in lifting water is costly. An adult male using efficient manual pumps takes about five days to lift as much water as can be lifted by one liter of fuel in a small diesel engine. An ox can lift three to five times as much as an adult male in 1 to 2 days of work, or as much as a liter of diesel fuel. Unless the marginal productivity of the water is extremely high, as it is in temporary droughts, more calories can be used up in human or animal energy than are produced by irrigating the crop.

Small, farmer-owned pump systems often evolve into a market system where pump owners use the surplus capacity of their pumps to water their neighbors' crops. With lifts (total dynamic head) of around 10 meters, areas throughout the Asian subcontinent, from Pakistan to India to Bangladesh, typically sell water for one-quarter to one-third of the crop harvest. Bangladesh has been particularly innovative in promoting pump irrigation systems for small farmers, who are able to rent pumpsets from state agencies on a seasonal or annual basis. Also, nongovernment organizations have helped groups of landless laborers purchase portable pumpsets, usually mounted on an ox cart. The groups then pump water for farmers on a rental basis. Efforts are now being made to encourage private enterprises in Bangladesh to provide rental pump services—like household equipment rental shops in the United States—perhaps under franchise arrangements. These companies may also rent small tractors and other equipment to farmers.

Alternatively, NGOs may develop and manage the pump irrigation systems for farmers. One of the more successful systems patterned after this model can be seen in Maharashtra, India. There, government agencies and sugarcane factories finance NGOs to oversee the development of large-scale pump irrigation systems covering several hundred hectares of land. The NGO builds and operates the system and collects water charges from farmers for debt service, operation, and maintenance expenses. One of the keys to the success of this system is an assured supply of low-cost electricity for the pumps.

Pump irrigation systems work well with PVC pipe conveyance systems and with sprinkler and drop systems that require pump pressure. Sprinkler and drip systems must have high-quality, clean water, preferably from aquifers. These systems are best suited for high-valued crops, where water supplies are scarce, the land is hilly and covered with pervious soils, and where there are labor and management constraints. For larger farms in the range of 125 to 300 acres, center pivot irrigation may be suitable. A private sector/IFC feasibility study in Kenya has found that it would be feasible to install more than 20,000 acres of center pivot irrigation for maize production there. With reasonably planned crops systems, center pivots could also provide many small farmers with controlled rain. All forms of sprinkler and drip systems, as well as pump systems, require an infrastructure of parts and service facilities, which makes them unsuitable for underdeveloped areas.

Engineering And Economic Aspects Of Pump Irrigation Systems

With lifts on the order of 10 meters and energy costs close to world market prices, pump irrigation is the most economical system. Few people are aware of this, because it is difficult to calculate the engineering and economic costs of pump irrigation systems under varying conditions. The results discussed here are drawn from a microcomputer-based model of pump irrigation systems applied to case studies in Senegal and Botswana [12] as well as in several Asian countries. The model can be used to compute irrigation requirements for various crop rotations, to size the pumpset, compute energy use, and calculate economic returns. As is almost always the case, the principal problem is to determine producer prices of inputs and outputs and expected increases in production under irrigated conditions. Although the figures used in the following case studies are the best that could be obtained under the circumstances, they have yet to be subjected to detailed review and revision.

The Case Of Senegal

Dispersed diesel pumpsets. Lift irrigation in Senegal was studied by Perlack and Petrich.[13] The irrigation system pumps water from the Senegal River to irrigate two crops per annum--sorghum and rice. The gross irrigated area is 800 hectares. Sixteen pumpsets are used, each of which irrigates a net area of 25 hectares, hereafter referred to as the unit command area (UCA). Three energy systems for irrigating this area were evaluated: (1) dispersed 25-kilowatt diesel pumpsets; (2) electric pumpsets, with energy provided through a "minigrid" from two 180-kilowatt diesel generating sets; and (3) two large diesel pumpsets.

Table 3.2 gives an example of the summary (the "driver") pages from the irrigation energy model. The study by Perlack and Petrich concentrates on the costs of the irrigation systems, and thus provides no information on the agricultural costs and returns. The figures for "net crop revenue" in Table 3.2 are therefore based on rough FAO [14] estimates of returns of US$300 to $400 per hectare per annum (with two crops a year) in SSA. As shown in section B of table 3.2, the model is used to calculate the annual energy costs, the present value of the total costs per hectare of gross irrigated area (GIRRA), the benefit-cost ratio, and the internal rate of return of the system over a life of 25 years. This period was chosen so that lift irrigation systems could be better compared with gravity systems.

At present, the total costs of this system amount to US$1,604 per hectare of GIRRA. This is only about half the cost of a medium-size gravity irrigation system in India, and an even smaller percentage of a typical gravity system in Africa. A benefit-cost ratio of 1.13 and an internal rate of return of 19 percent make this project economically feasible.

Large diesel pumpsets. Simulations indicate that the electrical "minigrid" system is not economically feasible. The system might work with fewer but larger diesel pumpsets, but this would depend on the layout of the country and organizational and managerial factors. If net crop revenue remains the same, the economic benefits of a large lift irrigation system

---

[12] David Seckler, R.K.Sampath, and S.K.Raheja, "An Index for Measuring the Performance of Irrigation Management Systems with an Application," Water Resources Bulletin vol.24, no.4 (Aug.'88),pp.855-860;David Seckler, Jack Keller, David Molden, David Garland, & Tom Sheng Design and Evaluation of Alternative Lift Irrigation Systems for Developing Countries, Report to USAID Office of Energy, S & T Water Management Synthesis Project (May 16, 1988).

[13] Robert P. Perlack and Carl H. Petrich, "A Comparison of Different Power Sources for Irrigation Pumping Systems in Two Agroclimatic Regions of Africa," Report to USAID, Bureau for Science and Technology (Oak Ridge, Tenn., 1987).

[14] Food and Agriculture Organization, Irrigation and Water Resources Potential for Africa (Rome, 1987); Irrigation in Africa South of the Sahara with Particular Reference to Investments for Food Production (Rome, 1987).

Table 3.2: COMPUTATION OF IRRIGATION ENERGY REQUIREMENTS

A. INPUT

1. Unit Command Area (ha) = 25
2. Crop System

	Crop 1	Crop 2	Crop 3	Crop 4	Crop 5
Crop:	Sorghum				Rice
% of total area:	100				100
3. Net crop revenue: ($ per ha)	200				200

	Input	Management Factor	Suggested
4. Equipment factors (refer to Table A)			
(a) D for diesel, O for other D			
(b) Engine/motor efficiency (%)	20	1	22
(c) Pump efficiency	60	1	70
(d) Engine life (hrs)	10000		
(e) Pump life (hrs)	15000		
(f) Conveyance System to Command Area			
C for Canal, P for Pipe P			
Length (m)	30		
Conveyance efficiency (%)	100		
(g) Within command area			
Conveyance efficiency in UCA (%)	100		
Application efficiency (%)	60		

	Input	Local Factor	Suggested
5. Capital cost			
(a) Engine/motor ($/Kw)	130	1	182
(b) Pump ($/Kw)	50	1	36
(c) Engine installation ($/kw)	50	1	36
(d) Pump installation ($/kw)	20	1	7
(e) Well miscellaneous	14000		
Conveyance system			
(f) Source to command area ($/m)	9		
(g) Within command area ($/ha)	20		

6. Energy factors
   (a) Cost of diesel ($/L) or   0.40
   (b) Other energy source ($/kWh)
   Diesel energy content (L/kWh)   0.096

7. Discount rate (%)   10

B. OUTPUT

	Crop 1	Crop 2	Crop 3	Crop 4	Crop 5
1. Irrigated Crop System	Sorghum				Rice
2. Area irrigated (ha)	25				25
3. Volume required (m^3/yr)	355708				175417
4. Energy required (KW hrs/yr)	89313				30396
5. Energy cost $	3421				1164
6. Crop Revenue $	5000				5000
7. Net irrigated revenue $	1579				3836
Per ha	63				153
8. Gross irrigated area (ha)				50	
9. Annual volume water required (m^3)			531125		
10. Annual energy required of engine (kWh)				119709	
11. Annual operating hours				1487	
12. Size of pump set (BP in KW)				18.2	
13. Energy costs per ha/pa GIIRA				92	
14. Total system variable cost per ha/pa GIIRA				196	
15. Present value total costs per GIIRA				1604	
16. Benefit/cost ratio				1.13	
17. Internal rate of return (IRR) (%)				19	

increase, because of economies of size in unit costs of equipment and the increased energy efficiency of the larger engines.

According to the model, two 140-kilowatt pumpsets would be needed to irrigate this area. The conveyance efficiency in the unit command area drops from 100 percent in the previous case to 80 percent, owing to the longer length of the conveyance system through earthen channels. The model could be used to explore the feasibility of lining canals or pipe conveyance systems to save on conveyance losses (and sprinklers to save on application losses), but this has not been done. The economic benefits of this model are much greater than the preceding one—the benefit-cost ratio increases to 2.38 and the internal rate of return to a fabulous 127 percent.

<u>Public utility company model: Large diesel pumpsets</u>. Although the potential returns to large pump systems are greater than they are for small systems, there are substantial problems in financing and managing these large systems and in delivering water supplies to large numbers of farmers. One solution might be to create a publicly regulated, private sector company to develop the irrigation system and to sell water to the farmers. Then the question would be how to charge for the water delivered to them. The best way would be to charge on a volumetric basis—that is, so much per hectare-meter, or acre-foot. However, unless there is a pipe conveyance system and the volume actually delivered per unit of time can be accurately measured, it will be impossible to prove how much water was delivered within a reasonable margin of error owing to variations in conveyance losses at different delivery points. A second-best alternative would be for the utility company to charge on a per crop per season basis. In this case, the charge would vary with the volumetric irrigation requirements of the crop. Farmers would therefore have an incentive to grow more water-efficient crops in relation to crop returns. This is the charge basis assumed here, although the price per unit of water can also be easily computed through the model.

It is assumed that the irrigation charge is US$75 per hectare for sorghum and $25 per hectare for rice (because sorghum requires more irrigation in the dry season). At these prices, the farmers would receive slightly higher net returns from their crops than they would under the dispersed pumpsets model, and the public utility would realize a benefit-cost ratio of 1.25, or an internal rate of return of 27 percent. Clearly, all parties would gain under this system.

This example shows that despite their managerial efficiency, small pumpsets leave a great deal to be desired in terms of technical efficiency. However, when pumping from aquifers, large pumpsets have substantial diseconomies of size in terms of higher lifts owing to the cone of depression of the water level in the aquifer.

<u>Alternative pump energy systems</u>. The model was also used to examine the possibility of replacing some of the expensive diesel fuel with producer gas made from wood chips or similar biomass materials. Although the technology and reliability of producer gasification is not dealt with here [15] it can reasonably be assumed that gasification would replace 65 percent of the diesel fuel at a conversion rate of 3.2 kilograms of air-dried wood per liter of fuel replace. The wood is assumed to cost $US50 per metric ton, delivered to the site and chopped into the appropriate sizes. This price is sufficient to yield a favorable return on

---

[15] See Gerald Foley and Geoffrey Barnard, "Biomass Gasification in Developing Countries," World Bank Technical Report no. 1 (Washington, D.C., January 1983); David Seckler, "Alternative Energy Technologies for Rural Areas in Developing Countries: A Cost-Effectiveness Analysis" (1984).

woodlots under most conditions. The gasification units cost $100 per kilowatt of capacity and can be operated for 20,000 hours with reasonable maintenance. In this application, the gasification unit yields a benefit-cost ratio of 1.55 and an internal rate of return of 18 percent.

The costs of solar cells appear to be decreasing to the point that they may be feasible sources of energy for pump irrigation in the next few years. The cost of electricity through a diesel genset is about $0.15 per kilowatt-hour. Electricity can now be produced by photovoltaics at a cost of $0.30 per kilowatt-hour, and new plants that are on stream may reduce costs to $0.15 per kilowatt-hour.[16] It is not clear whether these costs are for the total system, which costs about twice as much as the photovoltaic cells. If they are for the total system, this is a viable technology for supplemental pump irrigation in remote areas of developing countries. These units have a virtually maintenance-free life of 20 years. Thus a one-time subsidy of about half the capital costs may bring the costs of solar energy to a manageable level for private sector irrigation. Of course, both wood gasification and solar energy technologies become more economical when petroleum has to be imported, since the shadow price of foreign exchange is often high.

The Case Of Botswana

Diesel pumpsets. The model was also used to evaluate the potential returns of pump systems in Botswana. The study by Perlack and Petrich was based on small tubewells with unit command areas of 4 hectares and an operating head of only 3 meters (compared with 10 meters in Senegal). Since all of the comments in the Senegal case also apply here, we need only point out the high benefit-cost ratio of 4.37 and internal rate of return of 191 percent in Botswana.

In addition, Jones et al[17] studied a lift irrigation system in the Etsha area of Botswana, where water is pumped from a lagoon and conveyed through 1.5 kilometers of pipe, with a lift of only 1.28 meters. The soil has high infiltration rates, so a sprinkler system is used to apply the water to farmers' fields. It should be noted that net crop revenues are very high because prices for new development projects in remote areas are highly subsidized. Even so, the economic returns to this project are barely acceptable--the benefit-cost ratio is 1.07 and the internal rate of return is 13 percent. These results are mainly due to friction losses through the pipe and sprinkler system.

Thus, under favorable environmental conditions, properly designed pump irrigation systems appear to be economically sound technologies for irrigation in Africa. Biomass gasification might be especially appropriate since it promises high-valued use of underutilized land resources on tree plantations. If capital subsidies are available, it might even be possible to use solar energy, especially in remote areas.

---

[16] H. M. Hubbard, "Photovoltaics Today and Tomorrow," *Science* (April 21, 1989), pp. 297-303.

[17] R. B. Jones, K. Rashem, and David Molden, "An Assessment of Irrigation Potential in the Western Ngamiland Area from the Perspective of the Small Scale Farmer," Department of Agricultural Research Technical Report no. 2 (Botswana Ministry of Agriculture, November 1986).

The Case Of Nigeria And Mali

I recently completed World Bank irrigation assignments in Nigeria and Mali, and several aspects of this experience are relevant to this discussion.

<u>Surface and pump systems.</u> It was most gratifying to see the response of farmers in both Nigeria and Mali to well-managed irrigation systems with reliable supplies of water. Their record of success thus far shows that with good irrigation systems and supplies of other inputs farmers in SSA will do as well as the best in India or Indonesia.

One of the remarkable aspects of gravity irrigation in Nigeria is that farmers are paying the operation and maintenance costs. In several large systems, farmers are paying about US$20 per hectare per crop, with collection rates in the range of 70 to 100 percent. Also, the combination of more free markets and reduced food imports has resulted in higher domestic prices for agricultural products in both countries. Consequently, the rates of return for investment in pump irrigation and agriculture have been very high in these two countries and have greatly stimulated private investment in these sectors.

Tank Irrigation Systems

Tank irrigation systems consist of water impoundments in gullies near the base of small watersheds. Since tank systems depend on runoff for their supply of water, they are best suited to areas where precipitation is erratic and intense. The tanks capture the water during periods of high runoff and store it for irrigation during temporary droughts. One of the advantages of this system is that the tank may be filled and emptied several times during the year, and thus the irrigation supply may be several times the one-time storage capacity. For gravity irrigation, the tanks must be elevated above the command area, although they can be combined with pump systems. The basin must be deep enough and round enough to provide a reasonable ratio of impounded water to dam costs, achieve reasonable evaporation losses, and provide dead storage for sedimentation. The viability of tanks depends in part on the size of the sediment loads from the watersheds. In India, some small tanks fill up with sediment in only one year, whereas others take centuries.

The tank irrigation model discussed here is the Sikhomajri system, named after a small village in Haryana, India.[18] The tank has a storage capacity of about 45 acre-feet of water, and serves a net irrigated area of about 80 acres with supplemental irrigation. The major crops in the monsoon season from late June to October are maize, rice, and sugarcane. What is the major crop in the winter season.

The tank and irrigation system began operating in 1978. Although it was not monitored as well as it should have been, informal observations confirm that supplemental irrigation from the tank, together with other irrigation-induced inputs, increased average yields per season two- to threefold, and annual production four- to sixfold. Indeed, one of the major problems for villagers during the first year was how to store and market the produce, a situation they had never faced before.

Another problem for any system in which water is shared is how to allocate the water among users, especially in small villages where water allocation can stir up omnipresent social

---

[18] David Seckler, "Institutionalism and Agricultural Development in India," Journal of Economic Issues, vol. 20, no. 4 (December 1986).

tensions. In an area not far from Sukhomajri, a government-sponsored irrigation program had to be suspended because the allocation system led to murder and mayhem in the villages.[19]

In particular, the "have" and "have-not" status of citizens may be disturbed by irrigation. The real income of landowners can double or triple, while the income of landless people may remain almost the same. The problem was resolved in Sukhomajri by giving every family in the village a right to the same quantity of water in the tank, no matter how much, if any, land they owned. Those with more land that water and those with more water than land then were then forced to truck, barter, and exchange. In most cases, landowners would let landless people farm and irrigate part of their land on a crop-sharing basis in return for water and labor.

This institutional solution was made possible in large part by a technological solution to another problem. The command area is very hilly and almost impossible to irrigate by gravity conveyance systems. This made it necessary to bury the PVC pipe conveyance system and to supply pressure from the elevation of the tank (about 40 feet above the highest point of the command area). Wherever gravity pressure is sufficient, buried PVC pipe is by far the most cost-effective irrigation distribution system. It can run up hill and down, it does not waste agricultural land for channels, and maintenance costs are virtually zero. Although concrete pipes can also be used, they are a maintenance nightmare in irrigated areas because land subsidence causes the joints to crack. Even more important from an institutional point of view, the discharge form pipes can be measured and water usage gauged. It is easy to calculate the discharge from a pipe if one knows the pressure at the source and friction loss in the pipe. Thus every farmer can be assured of receiving the quantity of water to which they are entitled. Furthermore, since the pipes are buried, water stealing is virtually eliminated, even if the pipes are damaged.

During the initial stages of development, the Sukhomajri system was managed by an informal development organization in cooperation with a water users' association elected by the villagers. This organization represented a collaboration of the Central Soil and Water Conservation Research and Training Institute, the Haryana State Forestry Department, and the Ford Foundation. It financed and managed the construction of the irrigation facilities and helped the villagers establish a water users' association (WUA). Water rights were established and enforced and water duties were collected to cover operation and maintenance (not capital) costs. After about three years, the WUA assumed full control of the system, and up to now it has been working at least as well, if not even better than it did at the outset.

Once irrigation was established in Sukhomajri, other private sector activities followed. The villagers invested in milk buffalo to consume the extra fodder. They found this such a good business that they tripled the size of the herd, and now fodder has to be imported to the village. Also, the organization provided credit and transportation assistance for small rope-making enterprises that grew up around the grass produced on watersheds. The village bid for grass-cutting rights on forestland, and this yielded a substantial profit, which was then invested in threshing machines and a school building.

Tank systems are often considered the most desirable irrigation technology because they use local resources for construction, but this judgment must be qualified. First, the

---

[19] Douglas Merrey, "Irrigation and Honor: Cultural Impediments to the Improvement of Local Level Water Management in Punjab, Pakistan," Water Management Technical Report no. 53 (Fort Collins, Colo.: Consortium for International Development, December 1979).

design and construction of a small dam requires engineering expertise and supervision (piping problems are common in small dams). Second, a tremendous amount of earth has to be moved even for small dams and this must be accomplished in a short period of time between rainy seasons. Few villages have the labor or the skills for this highly intensive effort and usually rely on outside earth-moving gangs and some machinery. Third, the real costs of the dam per unit of water storage are necessarily higher than for larger reservoirs because of physical economies of size. Some of this cost is offset by smaller distribution systems and better irrigation management. However, a small tank system is likely to cost about the same per unit of water delivered as a well-designed and constructed medium irrigation system with a command area of roughly 5,000 to 10,000 hectares. This would be about twice the present value of the total cost per irrigated hectare of pump systems. Tanks can also be used as "percolation" devices to recharge aquifers for pump irrigation. In areas with pervious soils, good aquifers, and high evaporation losses, conjunctive use of percolation tanks and pumps may be the best irrigation system.

Even though a small tank system costs about the same as a medium surface system, it is usually more flexible and provides better control over water use. The "melons on the vine" model of China and the Mahaveli project in Sri Lanka provide good examples of how the advantages of both systems can be captured. In these systems, medium to large storage and diversion systems feed water into smaller, locally controlled tanks. The physical economies of size of water storage are combined with the managerial economies of size of small, locally controlled systems.

Tanks have an advantage over pump systems in remote areas where parts and repair services are not available. In Africa, where pump equipment must be imported, tanks also provide a means of decreasing the foreign exchange cost of irrigation. And, like pump systems, they do not require highly sophisticated and costly engineering expertise for design and construction. Furthermore, when tanks can be erected high above the command area, they provide gravity pressure through pipes to sprinkler and drip systems. The same technique of capturing natural head through pipes can be used in high streams, simply by diverting water into the pipe. Where these opportunities exist, they provide perhaps the best of all irrigation technologies. Several such units have been installed in Guatemala with highly satisfactory results.[20]

Although I believe that pumps and tanks should be the lead irrigation technologies in Africa, I do not wish to close the door on large-scale storage and diversion projects. In very arid areas that require full irrigation, these systems offer the only means of increasing food production to a substantial degree. People who criticize these and other irrigation systems in Africa fail to provide a rational alternative to irrigation.[21] With population doubling every 25 years or so, the arid areas of Africa are condemned to increasing food problems, rural poverty, and lost opportunities for foreign exchange earnings through agricultural imports if they do not adopt large irrigation systems. The best advice to give countries looking to large

---

[20] Alan D. LeBaron, Tom Tenney, Bryant D. Smith, Bertis L. Embry, and Sandra Tenney, *Experiment in Small Scale Sprinkler System Development in Guatemala: An Evaluation of Program Benefits*, W.S. Project Report 68 (Logan, Utah: Utah State University, 1987).

[21] Jon Moris, "Irrigation as a Privileged Solution in African Development," *Development Policy Review* (London, 1987).

surface systems is to study and adopt the major features of the [22] warabandi system of northwest India.

Intermediary Agencies For Private Sector Irrigation in Africa

Wesley C. Mitchell, one of the great American economists, once observed that economic texts are part microeconomics and part macroeconomics, and the only thing that holds them together is the binding! The more I think about private sector irrigation the more I believe that what we need to better understand is the "mesoeconomics," that is, the institutions between the macro and micro, between the purely public and the purely private sectors.

To describe the function of these mesoeconomic institutions in irrigation we may draw on a fundamental theorem in agricultural development formulated by another great American economist, Theodore W. Schultz. Schultz contends that farmers in developing countries are "poor but rational." By this he means that, given their resources and environment, poor farmers are already at an economic optimum, and that the only way their situation can be changed is to provide them with a "new input" that will enable them to move to a higher optimum. Among the most important new inputs are those that substantially change the operating environment of farmers. These inputs do not appear by themselves. Nor is it profitable for private sector enterprises to invest in other inputs until the environment has been changed. It is a major function of mesoeconomic institutions to help create the environments in which the private sector can flourish.

The Sukhomajri program mentioned earlier provides one of many examples of the important role of such institutions. In addition to providing the physical infrastructure for irrigation, the development organization there helped the WUA enforce water allocation rules while they were becoming institutionalized. It also helped repair and maintain the irrigation system. One of its most important functions was to absorb the costs incurred in learning by doing. For example, it covered the cost of an elegant and expensive sprinkler system installed on the higher land in Sukhomajri, which for reasons that are still not clear, was not used by the farmers and had to be abandoned. I believe that intermediary organizations of this kind could work between government agencies and private enterprises to support the development of private sector irrigation in Africa.

Intermediary organizations fall into four main types, grouped by their profit status. The nonprofit category contains private voluntary organizations (PVOs) and cooperatives; and the for-profit groups are public utility companies and agricultural development corporations.

Nonprofit organizations. Many PVOs have come to regard their traditional charitable orientation as inadequate. They find that in order to do good they must acquire the technical and managerial expertise needed to oversee development. Meanwhile, cooperatives have found that they cannot simply provide marketing services to their farm constituents, but must also help them change the environment if development is to proceed satisfactorily. Properly managed, technically competent, and politically independent PVOs and cooperatives are able to provide a bridge between the purely public and private sectors because of their nonprofit status; although they are supposed to help others make profit, they themselves cannot do so. Thus their use of public funds does not have to be policed as much and they have

---

[22] S. P. Malhotra, *The Warabandi System and Its Infrastructure* (New Delhi, India: Central Board of Irrigation and Power, 1982); Seckler et al., "An Index," and *Design and Evaluation*.

greater flexibility in using these funds to generate profitable private sector activities. Nonprofit organizations can help to establish enterprises that can be handed over to for-profit organizations, when they become profitable. In order to achieve this objective, the enterprises must be designed and operated from the start to achieve a commercial status in the future.

For-profit organizations. A public utility company is a privately owned and operated profit-seeking company that has a natural monopoly position and is therefore subject to government regulations pertaining to price, quality of service, and return on investment. Natural gas, electricity, and telephone companies in the United States are typical examples of public utilities companies. A public utility company for irrigation would sell water to farmers on a commercial basis. The company would finance, own, and operate the system. In principle, a public utility company can handle any kind of irrigation technology. Large publicly owned and operated surface irrigation systems could be privatized to a public utility. The viability of this model depends on economics (which may be improved by subsidies on capital costs), effective management, the ability to collect water charges from farmers, and on a well-functioning public utility agency. Perhaps this is why public utilities have failed to move into irrigation. However, if the economics are right, there is no reason in principle why this model would not work in irrigation as it does in other public goods.

Agricultural development corporations (ADCs) are large agricultural businesses that own, lease, or manage large areas of land for purposes of agricultural development. In principle, ADCs are capable of providing any form of irrigation for themselves. There are many large ADCs in Africa. Indeed, the operations of Lonrho Ltd. in Africa alone make it one of the largest commercial agricultural operations in the world. Although many people balk at the idea of very large ADCs in Africa, and these are highly risky ventures for investors even under the best conditions, they may offer an important institutional mechanism for irrigation and agricultural development in some areas, especially if the emphasis is on development. ADCs may develop an area, operate it to establish a viable system, and then spin the land off to smallholders. Much of the agricultural area of the western United States was developed in this way.

The ADC model also is reflected in the "nucleus-smallholder" estate program of Indonesia and other countries of Southeast Asia. Although this concept is applied mainly to perennial crops to deal with the long-gestation periods and postharvest processing and marketing, it can also be used to solve other agricultural problems of development. The ADC model in this form would evolve through three stages:

(1) Development. A large tract of land with development potential would be sold or leased to the ADC. The ADC would then develop irrigation facilities; prepare the land; build roads, warehouses, and processing facilities; and farm the land with hired labor.

(2) Establishment. Selected laborers would be assigned viable areas of land to farm as tenants, and the ADC would provide them with inputs and output markets under favorable conditions.

(3) Devolution. After a few years, land titles would be sold to successful tenants on favorable terms. The ADC might retain certain "nucleus" activities, like input/output marketing. It might also continue to operate the irrigation systems in the public utility mode. Or it might sell the facilities altogether, perhaps to a cooperative of

the farmers, a public utility, or to a private corporation, leaving successful development behind.

Clearly, many financial, legal, and other details would have to be worked out for this rendition of the ADC model. But there is no reason why it could not work under the appropriate institutional conditions.

In sum, there are several kinds of intermediary organizations to support irrigation projects, both in the nonprofit and for-profit modes. The more activities that can be transferred to these organizations, the more that public agencies can concentrate on their proper role, which is to act as investors and entrepreneurs, rather than as managers of development. These organizations could oversee turnkey irrigation and agricultural development projects on behalf of host governments and donor agencies under carefully spelled-out payment and performance conditions. This is an area in which the International Finance Corporation has developed expertise through its experience with the private sector in other aspects of development. After the projects have been judged technologically, economically, and institutionally sustainable, the intermediary organizations would spin them off and move on to develop other demonstration projects. I believe that this model of government agencies acting in an entrepreneurial, venture capital role, with intermediary organizations implementing demonstration projects to be replicated by the private sector, could become a new and effective model for irrigation and other development programs in the future.

Irrigation Policy And Evaluation

Another important aspect of irrigation is the policy environment. Economists generally consider policy analysis to be their exclusive area of expertise. Policy is indeed the front line of the "imperial science." However, many outside the discipline believe that "noneconomic" criteria should be used to evaluate policies governing issues that relate to the long-term welfare of nations. For example, some critics contend that irrigation projects should not be selected solely on the basis of their projected internal rates of return, benefit-cost ratios, or other strictly financial criteria. I agree with this opinion in principle. The problem is not with economics, but with those economists who have neglected the fundamental principles of "welfare economics" in performing their evaluations. Welfare economics is the source and inspiration for the entire discipline. All too often it is ignored in financial evaluations.

This omission is particularly odd because one of the world's leading welfare economists has written a large volume on how to use welfare economics in project evaluations.[23] This book illustrates the appropriate ways of calculating internal rates of return (IRRs). Despite these guidelines, many economists continue to calculate IRRs incorrectly. (It is precisely because the IRR is calculated improperly that it conflicts with the benefit-cost ratio presented above.)

Even more important, irrigation projects should be evaluated on the basis of consumer and producer surpluses, not revenue. Again, nobody follows these principles. As Mishan says, "The consumer's surplus is the most crucial concept in the measurement of social benefits in any social cost-benefit calculation."[24] However, "consumer's surplus" does not appear in

---

[23] E. J. Mishan, <u>Cost-Benefit Analysis</u>, rev. ed. (New York: Praeger, 1976).

[24] Ibid., p. 24.

the table of contents or index of the work most commonly used to guide project evaluations.[25]

Welfare economics explicitly recognizes that economists must play a subsidiary role in the formulation of economic policies because policy is ultimately determined by social and political values. Economists are no more qualified (often they are less qualified) to make value judgments than the man on the street. As Oscar Wilde might have said, economists know the price of everything but the value of nothing. Such concepts as "economic efficiency" are meaningless except in relation to independently determined social objectives. Before policy analysis can be performed, the social objectives must be clearly defined and agreed upon. To my knowledge, every country in the world that has a substantial population and agricultural sector pursues basically the same set of social objectives, which may be grouped under the principle of "food security":

- To provide a secure supply of low-cost food for their people

- To maintain a reasonable degree of self-reliance in food production (for example, a balance of trade on the agricultural account)

- To achieve a high level of rural employment.

Some economists criticize these objectives on the grounds of economic inefficiency, especially the alleged gains from pursuing comparative advantage in international trade. However, both the concepts of comparative advantage and of economic efficiency become highly ambiguous in developing country settings. Their meaning is reasonably clear when physical and human resources are fully developed and employed—that is, when a country is "developed." When resources are underdeveloped and underemployed, the task is to develop and employ these resources—even if by inefficient means, whatever that may mean in this context.

In any case, I believe that the principle of food security provides quite a reasonable set of objectives for developing countries to pursue under most conditions. Low-cost food contributes to political stability (a necessary condition of economic development) and keeps real wage bills down, thereby enhancing comparative advantages in international trade. A reasonable degree of food self-reliance conserves foreign exchange, which may then be used to import productive items that cannot be made at home. Rural employment improves the welfare of rural people, who constitute the vast majority of the population, and of the poor in most developing countries. Rural employment also provide a means of rationally phasing the transition to the predominantly urban, industrial economies of the future, thereby reducing the appalling social costs of such market-driven transitions as have occurred in the United States.

Even if some economists do not agree with the principle of food security in general, it is difficult to see how they can argue against it in Africa countries. Many developing countries in Africa and elsewhere are experiencing a profound regime switch because of their debt crises. These countries no longer have the foreign exchange to pay for food imports. They either have to produce more food for themselves, or become increasingly dependent on unreliable food aid, or suffer food shortages, political instability, and reduced economic growth in all sectors.

---

[25] J. Price Gittinger, Economic Analysis of Agricultural Projects, 2d ed. (Baltimore, Md.: Johns Hopkins University Press, 1982).

Of course, this is not to deny that export crops can earn foreign exchange, which can be used to purchase food corps that cannot be economically grown at home. Export crops are also an important means of earning foreign exchange for fertilizers and other imports needed for local food production. It is as important to avoid food fundamentalism in agricultural policies as it is to recognize the great difficulties and risks in excessive reliance on agricultural exports in intensely competitive and highly subsidized international markets.

The principle of food security has important implications for evaluating projects in agriculture and irrigation. First, agricultural projects should not be evaluated solely on the basis of border prices, which largely reflect the agricultural subsidy programs of other, especially developed, countries. Second, project evaluations should include revised estimates of the shadow prices of foreign exchange, which are rapidly increasing with the debt crisis. Third, the shadow price of labor and other local factors should reflect both underemployment and the benefits of developing human capital through learning by doing. Fourth, shadow prices should be estimated on the basis of long-run conditions of supply, demand, and prices at the sectoral, rather than the project level. In the case of irrigation, projects are evaluated under the assumption that they will not affect output prices. Although this approach may be valid for individual projects, it does not necessarily hold for irrigation projects taken as a whole, and can give rise to a composition fallacy.

It may be found that no single irrigation project will yield an acceptable return, and therefore that all projects will be rejected. However, if there are no irrigation projects at all, rapid population growth is likely to cause output prices to increase. With increased prices, the rejected projects might have had favorable returns, and in retrospect should have been implemented. In such cases, the proper course of action is to first perform a sectoral evaluation on the basis of the effects of differing quantities of irrigated areas in a country on agricultural prices. Then individual projects can be selected on the basis of the resulting schedule of projected prices down to the appropriate cutoff point.

Since one of the objectives of food security is to keep food prices from rising too rapidly, prices and revenues are not the proper criteria by which to evaluate benefits. Instead, benefits should be judged on the basis of consumer and producer surpluses, and the socially desirable distribution of the two. For these and other reasons, I believe that the proper economic evaluations of well-designed and implemented irrigation projects would yield much higher social rates of return in Africa than is commonly thought. Once social returns have been properly estimated, public sector programs and private sector prices can be brought into line with social objectives.

Map 3.1

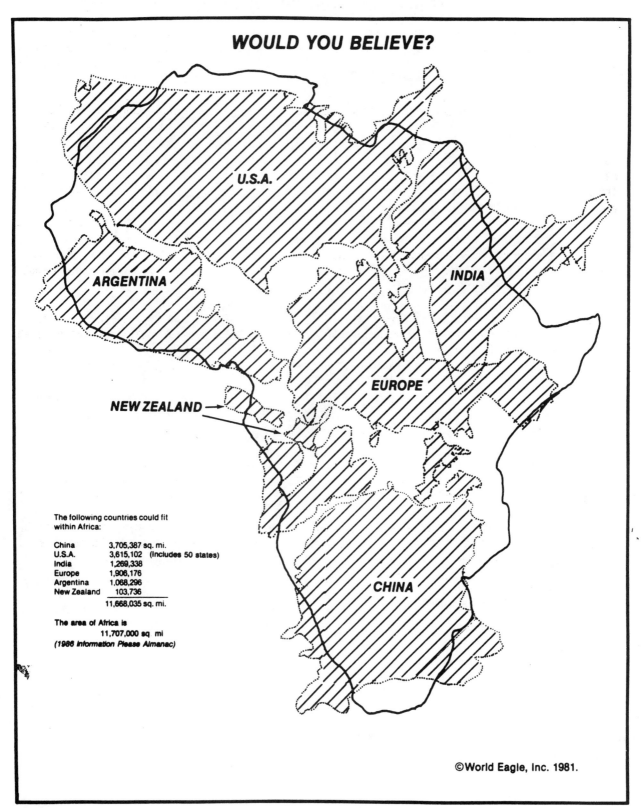

Map 3.2

MAIZE: RAINFED PRODUCTION

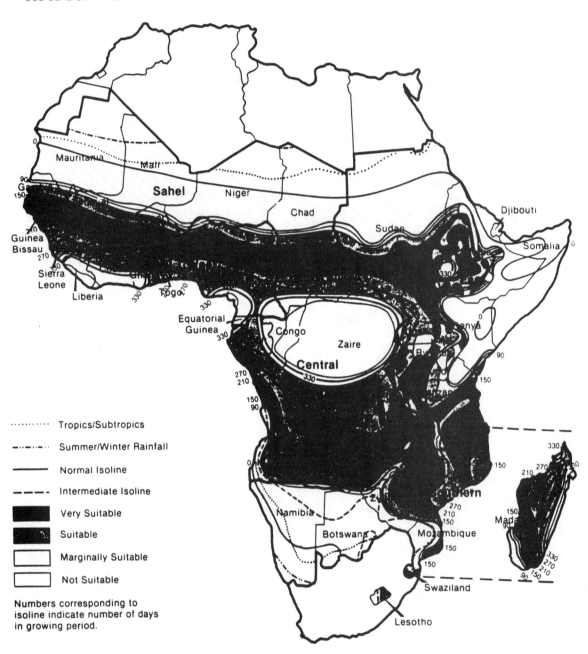

Source: U.S. Department of Agriculture. Foreign Agricultural Research Report No. 166. *Food Problems and Prospects in Sub-Saharan Africa: The Decade of the 1980's*.

South Africa obviously is part of Sub-Saharan Africa. Unfortunately comparable information was not available to add to this map created by the U.S. Department of Agriculture.

Map 3.3

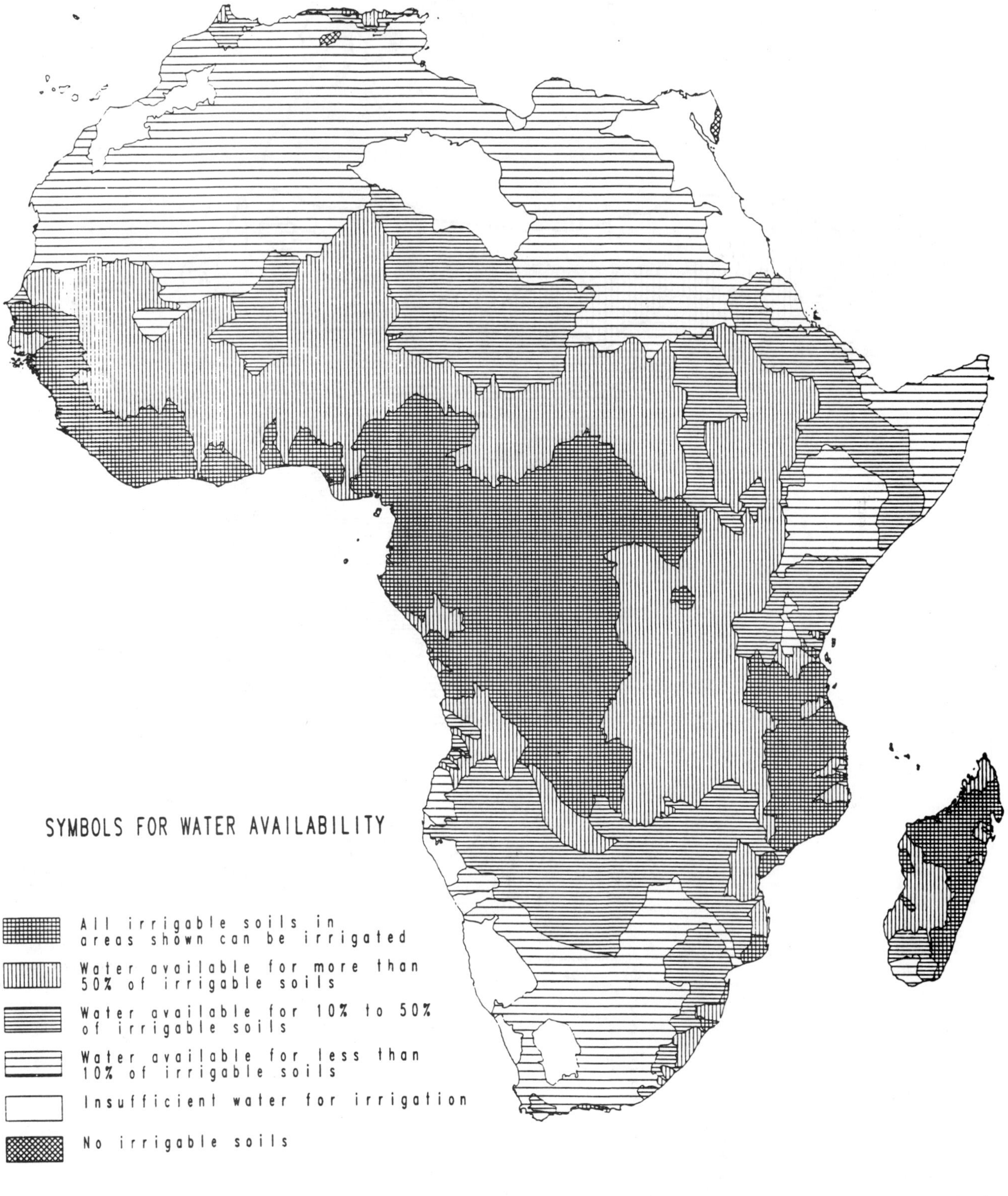

SYMBOLS FOR WATER AVAILABILITY

- All irrigable soils in areas shown can be irrigated
- Water available for more than 50% of irrigable soils
- Water available for 10% to 50% of irrigable soils
- Water available for less than 10% of irrigable soils
- Insufficient water for irrigation
- No irrigable soils

(From FAO, 1987 Irrigation and Water Resource Potential)

Map 3.4

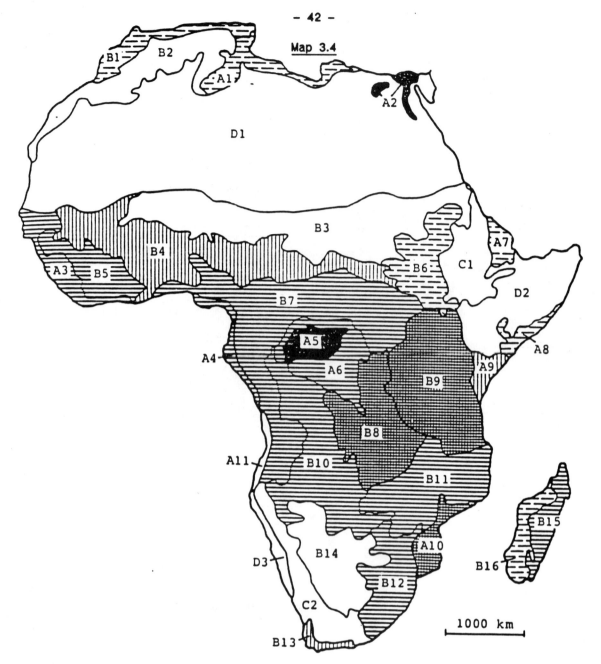

	SDF	WDF		SDF	WDF		SDF	WDF		SDF	WDF
A1	.7	.2	B1	.7	.2	C1	.7	.5	D1	-	-
A2	.9	0	B2	.4	.1	C2	.7	0	D2	-	-
A3	.5	.6	B3	.5	0				D3	-	-
A4	.6	.8	B4	.6	.3						
A5	.8	.9	B5	.6	.6						
A6	.5	.8	B6	.5	.1						
A7	.4	.1	B7	.5	.7						
A8	.6	.2	B8	.6	.7						
A9	.7	.4	B9	.6	.6						
A10	.7	.6	B10	.5	.5						
A11	.5	.1	B11	.6	.5						
			B12	.6	.6						
			B13	.6	.4						
			B14	.5	.1						
			B15	.5	.7						
			B16	.8	.2						

Map 3.5

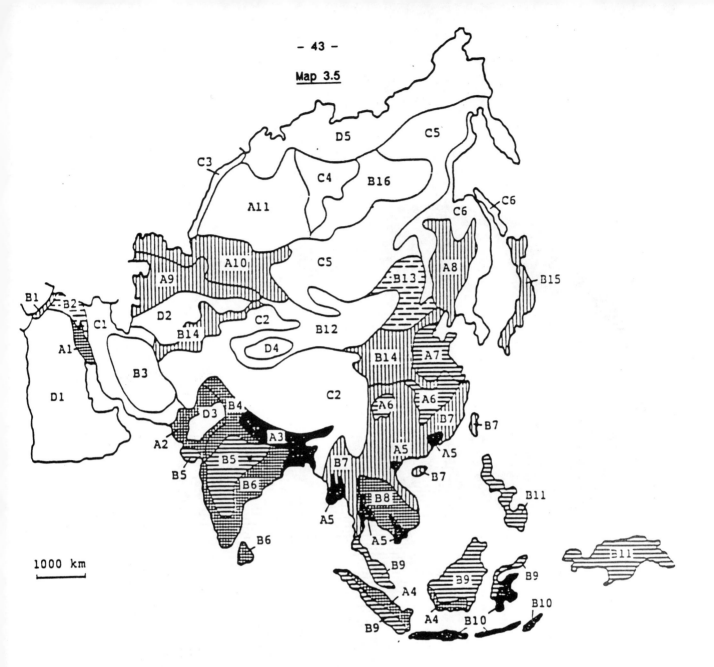

	SDF	WDF		SDF	WDF		SDF	WDF		SDF	WDF
A1	.8	0	B1	.8	.2	C1	.6	.1	D1	-	-
A2	.7	0	B2	.7	.1	C2	.6	.1	D2	-	-
A3	.9	.7	B3	.6	0	C3	.6	.9	D3	-	-
A4	.6	.9	B4	.7	.3	C4	.6	0	D4	-	-
A5	.9	.8	B5	.5	.3	C5	.6	.9	D5	-	-
A6	.8	.9	B6	.6	.6	C6	.6	.9			
A7	.9	.5	B7	.5	.9						
A8	.9	.7	B8	.7	.6						
A9	.6	0	B9	.5	.9						
A10	.9	.6	B10	.7	.8						
A11	.5	.9	B11	.5	.8						
			B12	.6	0						
			B13	.7	.3						
			B14	.8	.5						
			B15	.7	1.0						
			B16	.4	.9						

Map 3.6

Major Areas for Irrigation Development

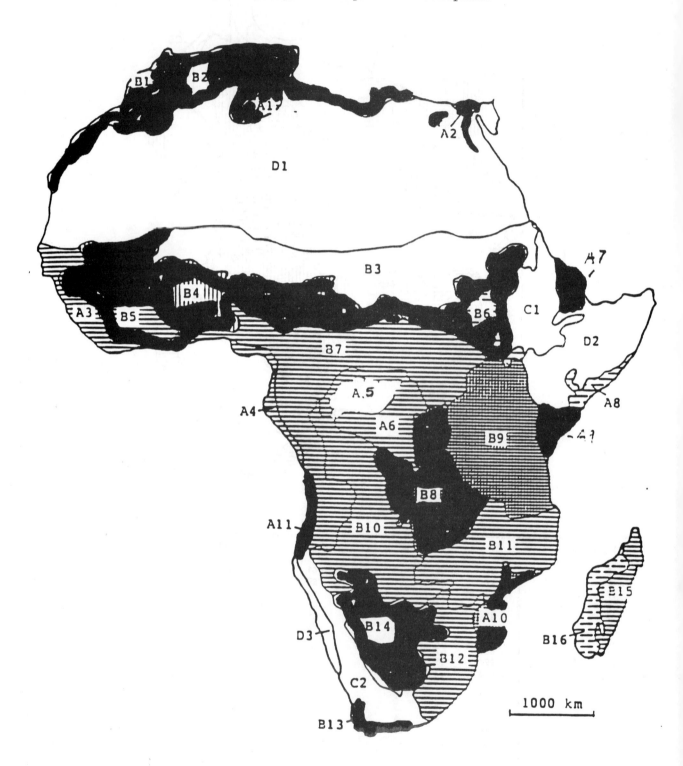

CHAPTER 4: HOW RISKY IS IRRIGATION DEVELOPMENT IN SUB-SAHARAN AFRICA?

Guy Le Moigne and Shawki Barghouti
(Mr. Le Moigne is Senior Adviser, Agriculture and Water Resources, and Mr. Barghouti is Division Chief, Agriculture Production and Services, both of Agriculture and Rural Development Department, PPR, World Bank)

During the past few decades large-scale irrigation schemes in Sub-Saharan Africa (SSA) have run into a number of problems, which have shaken the confidence of potential investors. Thus it has become important to identify and, where possible, reduce the economic and technical risks associated with this type of investment. This chapter examines the risks and presents some thoughts on how to develop a strategy for advancing irrigation in this zone to sustain agriculture despite years of drought.

A Review Of Irrigation In Sub-Saharan Africa

The farming systems in SSA have evolved largely in response to the availability of water, which is irregular at best. As a result, some form of traditional irrigation has been practiced in this region for centuries. Perhaps the oldest system is flood irrigation, whereby crops are grown in naturally inundated areas, as is the case in Sierra Leone and Gambia, where rice is produced in natural swamps. This system is used on floodplains and on the banks of rivers and seasonal streams throughout Africa, especially in Niger, Mali, Sudan, Somalia, and Northern Nigeria. Farmers also discovered long ago how to construct barriers to retain water and gradually developed water techniques (diking, desalination, transplanting) to suit the conditions of traditional irrigation. In addition, they used various devices (e.g., the sakyah in Sudan, the shadouf in Nigeria and Niger) to lift water from seasonal streams or rivers in order to irrigate areas of about 1 to 20 hectares for family food and cash crops.

Systematic or modern large-scale irrigation was not introduced in SSA until the early 1900s, which is well over a century later than the time of its arrival in India, for example. One of the first large systematic projects was the Office du Niger, in Mali, which was surveyed in 1919 and launched in 1925. The main works were not begun until 1934, however, and the project was not brought into full service until 1948. The objective was to support cotton monoculture. Although the project was designed and constructed satisfactorily, it met with some serious problems. It is now devoted to rice cultivation, but the yields are poor. Large-scale systematic irrigation is not practiced extensively, except in Sudan and Madagascar. The Gezira scheme in Sudan, which irrigates about 950,000 hectares under one central management, also started in 1920. More recently (in 1974), the Rahad scheme in Sudan was built after the Gezira model to irrigate about 150,000 hectares. International agencies have financed several projects designed to help develop modern irrigation schemes in SSA, and have supported small-scale projects, particularly in the drought-stricken areas of Chad, Mauritania, Senegal, and Burkina Faso.

Despite these efforts, irrigated agriculture has had limited success in SSA, and in 1989 it appears that almost all of the large irrigation schemes are experiencing some difficulty with management and operation. The performance of small-scale schemes has been better, but still mixed. Consequently, irrigation development has come to be regarded as a risky investment, and whatever potential exists for further development has been slow to materialize, even though the results of the large schemes undertaken in Sudan (Gezira and

Table 4.1: COMPARISON OF BASIC DATA ON IRRIGATION PROJECTS FROM A FEW SELECTED COUNTRIES SOUTH OF THE SAHARA

COUNTRY Name of Project and Closing Date	Cost of Irrigation Works (US$ M)	Bank/IDA Lending (US$ M)	Total Project Cost (US$ M)	Implement. Period		ERR Expected (%)	ERR Actual (%)
				Expected (in yrs)	Actual (in yrs)		
BURKINA FASO							
Second Rural Development Fund (April 1982)	6.04	9.4	18.5	7	7	16	12
CAMEROON							
Semry Rice I (June 1976)	3.55	3.7	9.3	4	4	13	23
Semry Rice II (September 1984)	21.6	29.0	55.5	6	8	15	20
KENYA							
Bura Irrigation (June 1986)	16.0	40.0	105.0	6	8	13	*
MADAGASCAR							
Lake Alaotra Irrigation (June 1975)	5.18	5.2	8.96	5.5	4.5	11	22 1/ * 2/
Morondava Irrigation (December 198)	n.a.	14.1	56.6	5	8.5	16	3/
MALI							
Mopti Rice I (November 1978)	5.62	9.5	13.1	6	6.4	14	17
Mopti Rice II (June 1985)	16.2	14.8	31.5	5	7	18	3
MAURITANIA							
Gorgol - Noir Irrigation (March 1982)	45.2	15.0	93.2	6	6	7.4	2.7
NIGER							
Irrigation I (June 1984)	18.4	15.0	24.5	4	4.6	12.5	3
NIGERIA							
Agriculture Rice Dev. I	24.3	16.3	46.6	5	6	17.6	*
Bida Agriculture Deve. (June 1986)	8.2	23.0	64.4	6	7	16	9.3
SENEGAL							
Debi-Lampsar Irrigation (March 1986)	22.62	20.0	36.7	5	8	10	-4
SOMALIA							
Northwest Agriculture (June 1985)	8.69	10.5	14.9	6.5	7.5	11	-2
SUDAN							
Rahad Irrigation (December 1982)	164.6	15.6	395.6	6	7.5	15	20

* Negative.
1/ 1977.
2/ 1981.
3/ Negative to zero.

Rahad), Cameroon (SEMRY), and Senegal (SAED) have been encouragir able 5.1). The high yields achieved for cotton, rice, and other crops indicate that sucl are technically and economically viable. As management and maintenance have det however, these schemes have produced fewer benefits in recent years.

Another problem for irrigation in SSA has been the controversy surrounding the appropriate scale for systems in this region. Some experts have argued that only large-scale irrigation can make an appreciable contribution here, despite the difficulties in operating and maintaining such schemes.[26] Small-scale irrigation, according to this argument, is only viable in countries with abundant land resources and an adequate supply of labor. Furthermore, systematic irrigation has had too short a history in SSA to prove itself. Thus it is not surprising that farmers and governments have not yet mastered the intensive form of crop production that irrigation demands. To complicate matters, there is not enough population pressure on the land when rains are favorable to create much interest in adopting irrigation. Some cultivators tend to neglect or even abandon small irrigation projects when the rains are good.

Erratic Rainfall Regimes

Many countries in this zone—such as Senegal, Gambia, Burkina Faso, Sudan, and Niger—receive about 50 to 800 centimeters of rainfall during the short rainy season. In some cases, more than half of the annual rainfall is received during a three-month period. Rainfed agriculture is unlikely to provide food security under such circumstances, which could, however, be alleviated by irrigation. Wherever population is on the rise and droughts are frequent, it takes more than better management of rainfed crops to meet food needs. Such regions must also invest in irrigation and put more effort into managing their soil and water resources. Otherwise they will be unable to fend off the debilitating consequences of drought cycles.

Irrigation has already done much to increase and stabilize agricultural production, even in the SSA countries where rainfall is adequate. Many parts of Ghana, for example, experience water deficits every crop season. Although Ghana's total rainfall is above average for SSA, it is highly irregular in several parts of the country and the distribution of rainfall in the country as a whole cannot support year-round short-term crops. Annual evaporation may be as high as 1,800 millimeters in areas such as the Northern Savannah, where the supply of rainfall water is already meager. To adjust to this situation, farmers tend to grow short-duration grains and vegetables, especially sorghum and millet. Water losses due to evaporation and runoff make it difficult to expand the growing season much beyond three or four months in the Sahel zone. Other dry areas of SSA are able to grow annual crops for only six to eight months of the year. In parts of the subhumid zones of Ghana, Niger, and Cameroon, the water stress period lasts as long as five months.

On average, many SSA countries experience a complete crop failure once every three years. (For Ghana and other more humid countries, the rate is every five years.) Some areas of a country may suffer more than others, and conversely, some may have more success than others. As a result the human and livestock populations of these areas are forced to move about in search of food and water. The SSA countries have therefore come to believe that despite the high cost and the difficulty of implementing irrigation, it is needed to sustain

---

[26] W.R. Rangeley, "Irrigation in Arid Africa," paper presented at the World Bank Irrigation and Drainage Seminar, December 1985.

crops during periods of inadequate rain, as it will increase crop intensification (through two short-term crops per year) and will stabilize farming communities around reliable sources of water. Irrigation can play a particularly important role by supplementing rainfall with water pumped from rivers or from shallow groundwater.

Irrigated areas in SSA. The total area currently under irrigation in SSA amounts to about 5.4 million hectares, or 2 percent of the entire cropped area in this region. About 2.2 million hectares were developed by country governments, primarily under large-scale schemes. Most of this land is cultivated by tenants who lease small areas (5-10 hectares) to produce food and cash crops under the supervision of a government agency. Irrigation is handled by commercial operations on about 0.5 million hectares, and the remaining 2.7 million hectares are under traditional (flood, swamp, surface, and low-lift) irrigation developed largely without government support. In favorable years, the areas under traditional irrigation are able to expand significantly because ample water is available. In years of low flow, these areas receive considerably less water and their crops suffer. Yields drop and modern production methods can only be used to a limited extent.

Many modern irrigation projects in SSA have been plagued by long construction delays and high cost overruns, most of which can be traced to poor surveys, inadequate preparation, unrealistic assessments of benefits, and the lack of systematic planning and administration. In addition, operation and maintenance have declined owing to weak management and the lack of local funds, and farmers have failed to achieve much food security or a stable income (a decline in commodity prices in recent years has aggravated this situation). As a result, irrigated areas have been neglected, production has remained low, and yields have declined. Nevertheless, according to a recent FAO study, [27] irrigation is likely to remain an essential element of food strategy in eight SSA countries in the coming years (Senegal, Burkina Faso, Niger, Mauritania, Mali, Somalia, Botswana, and Kenya), which contain 14 percent of the SSA population. There is little chance for rainfed agriculture to expand in these areas while demographic pressure on rainfed land continues to rise. Although rainfed agriculture will no doubt remain the favored method of production in SSA countries, sizable areas in half of these countries are prone to drought, and small-scale irrigation employing small dams and groundwater could do much to reduce rural hardship.

The prospects for irrigation development in SSA. A striking feature of SSA is that its surface water is unevenly distributed. There are 24 large rivers in the region, but only the Senegal, Niger, Nile, Tana, Athi, Shebelli, and Juba rivers flow through the drought-prone areas. Although the total irrigation potential of the region is on the order of 33 million hectares, only 5.3 million hectares (about 16 percent) have been developed thus far.[28] In the eight countries cited above, about 36 percent of the irrigable land has been developed. (See Table 4.2 for the potential water resources of a few SSA countries.)

As already mentioned, half of the irrigated area in SSA is under modern large-scale schemes, many of which are experiencing problems of one kind or another, as illustrated by the various irrigation systems along the Shebelli River in Somalia. The crops produced there

---

[27] *Irrigation in Africa South of the Sahara*, FAO Investment Centre Technical Paper 5 (Rome, September 1986).

[28] The FAO has optimistically estimated the irrigation potential of Africa south of the Sahara to be 33.6 million hectares. More recent data for five countries in SSA indicate that the FAO figure is about 20-25 percent too high.

range from sugar, bananas, and citrus (all of which require year-round irrigation), to paddy rice (which requires flooded conditions for 80-90 days), to vegetable and upland rice (which needs generous supplies of water for up to 140 days), to maize and sesame (which are commonly grown under single "wild flood" irrigation). Thus, neither the irrigation nor the cropping system is uniform in this area, and there is no system for allocating and managing water. Before such problems can be solved, farmers will have to organize themselves into water distribution, drainage, and reclamation units, with a small but capable and motivated

Table 4.2: **COMPARISON OF WATER RESOURCES POTENTIAL OF A FEW AFRICAN COUNTRIES**

COUNTRY	AREA ($10^3 KM_2$)	WATER RESOURCES (KM^3)		
		GROUND	SURFACE	TOTAL
Benin	113	3.08	16.6	20.4
Botswana	601	12.2	0.7	12.9
Burkina Faso	274	4.3	22.8	27.1
Chad	1,284	46.7	36.4	83.1
Cameroon	475	15.4	117.3	132.7
Ethiopia	1,222	17.2	89.1	106.3
Kenya	583	9.1	14.8	23.9
Madagascar	587	15.3	115.0	130.3
Mali	1,240	8.9	31.1	40.0
Mauritania	1,031	2.0	0.0	2.0
Mauritius	1.9	0.05	0.8	0.8
Niger	1,267	10.9	2.1	13.0
Nigeria	924	49.3	145.4	194.7
Senegal	196	14.0	11.5	25.5
Somalia	638	6.0	0.5	6.5
Sudan	2,506	68.3	80.6	148.9

government irrigation management service. Without such a system, it will be difficult to improve crop production.[29] The prospects for irrigation can be advanced through the construction of water tanks and the development of groundwater resources.

Water tanks and reservoirs. In vast areas of SSA people still rely solely on the traditional system of constructing channels along contours to collect water in tanks and reservoirs, and intercepting desert streams (wadis) that carry water irregularly. Some recent improvements in traditional techniques have helped to increase efficiency. For example, farm communities have been encouraged to use pumps and to operate and maintain them, and also to collect dues from other water users of the same pump. Single pumps (on river banks or water wells) are owned and operated by individual families or small farming communities, depending on the local laws or customs. Owners (operators) have recently found it difficult to operate these pumps. Many fail to regularly service and maintain the pumps. Other operators at times find spare parts and fuel in short supply and services (logistic and repair) inadequate, especially in remote areas. In addition, they have to keep up with annual desilting and redredging chores in water tanks and reservoirs, and must also keep inlet and outlet

---

[29] Somalia Agricultural Sector Survey, main report and strategy (Washington, D.C.: World Bank, 1987).

channels clear. These reservoirs are fed either by intermittent streams or from their own depressions. Most of them go dry two or three months after the end of the rainy season.

Development planners tend to pay little attention to the design, construction, and maintenance of water tanks and field reservoirs. Consequently, few advances have been made in the systematic investigation or the development of these tanks under the soil, wind, and evaporation conditions of the SSA zone. Water tanks are usually constructed in a low-lying area or adjacent to a seasonal stream. Although they vary in size, most are excavated to a depth of 1 to 2 meters. Few reservoirs can supply water year-round, however, because of the high rate of evaporation, which is about 5 to 10 millimeters per day, or close to 2.5 meters over the dry season (which lasts 280-290 days). Another problem is seepage, which can reach 50 to 75 centimeters. Consequently, even if no water were used for domestic, livestock, or agricultural purposes, water losses would equal or exceed 2.5 meters. A minimum depth of 5 to 7 meters is therefore required if field reservoirs are to persist year-round.

Oil-based chemicals may be used to cover the water in the reservoir and thus reduce evaporation and water losses, while plastic covers can help reduce seepage, as long as the catchments are fenced and protected. With adequate treatment, catchment basins can yield more water. Although chemical treatment maximizes runoff, simpler techniques, such as vegetation control, are frequently more cost-effective. An alternative approach is to build stream barriers below the level of the stream bed. The resulting reservoirs will have only minimal evaporation losses.

Still another option is the horseshoe-shaped dam, which is particularly attractive because it can hold more water per excavated cubic meter than water catchments. A dam with suitable side walls extending upstream to hold 2 to 3 meters can provide depths of 5 to 6 meters at the spillway level, and the stored water could last year-round. The dam could be supplied with a siphon, sand filter, two low-lift high-capacity pumps (10 liters/second), fencing, and stock-watering troughs to serve community needs.

The financing of operation and maintenance and the method of recovering capital investments costs would have to be agreed with the users and farmers well before the dam was constructed. The maintenance of the dam, for example, could be financed from the proceeds of water purchased by the resident communities or pastoralists who visit the dam site occasionally to water their livestock. Farmers can manage this system if adequate assistance is provided.

Potential for groundwater irrigation. Although several agencies have collected information on groundwater in SSA, little has been done to consolidate it or apply it to systematic projects. According to a recent FAO study, groundwater appears to be more widespread than surface water in much of the Sahel, and is exploited mainly for domestic and livestock purposes. However, it is being obtained from traditional wells with yields that are too low for irrigation.

Several studies have attempted to asses how groundwater can be used for irrigation and rural development in SSA.[30] Some countries have recently begun to exploit aquifers, especially in central Senegal and the area south of the Niger River. The information on water availability obtained with improved drilling techniques suggests that there is considerable

---

[30] R. Gouzes, J. C. Legoueil, and F. LeLandier, Groundwater and Rural Development in Sub-Saharan Africa, BRGM, France, 1985.

potential for expanding groundwater development in the region. Another encouraging sign is the fast-developing technology for the use of solar energy, electrical generators, and diesel engines for lifting and distributing water (especially through plastic pipes, which will help reduce losses, or ensure more equitable distribution and a better delivery system, particularly in marginal soils).

Village water projects in many SSA countries (financed by local initiatives and supported by nongovernmental agencies) have demonstrated that groundwater could be efficiently exploited and managed at the local level. Wells have been dug by hand and boreholes drilled by simple techniques that allow water to be bailed or pumps installed with a small range of lift. Thousands of shallow wells with an average depth of 50 to 70 meters have been drilled in this way and are already supplying water for human and livestock needs. With further work, they could also be used for crop production, especially for supplementary irrigation in the dry year, or for small vegetable gardens.

Caution would have to be exercised when attempting to integrate water use for human consumption and crop production, as explained in a recent review of a rural development project in Cameroon designed to improve rural water supply and irrigation.[31] Under the Cameroon project, wells and boreholes were constructed, rehabilitated, and equipped, and four small dams were built. In addition, 300 hectares of bottomlands were developed for irrigated rice production and small-scale market gardening was established through the provision of loans for the purchase of 60 motor pumps.

The results of these efforts have been mixed. The dike built for bottomland irrigation has never held enough water, especially in years of drought. Without proper extension and research, vegetable production for local markets has been limited to onion production, which has exceeded the marketing capacity of the region. The project has also run into other problems owing to weak institutional support, poor marketing facilities, inadequate engineering, and the lack of cost recovery measures to finance proper operation and maintenance of the water pumps. However, the arrangement making it possible for farmers to have joint ownership of the motor pumps has operated efficiently, despite the lack of maintenance and poor supervision. Clearly, small-scale irrigation schemes are not problem-free, but they have an important role to play in improving agricultural production in SSA.

Some agricultural and water experts argue that investment in groundwater development cannot yet be fully endorsed in SSA because the information on the recharge rates in existing aquifers is still too meager. Thus the water supply and sustainability of water yields remain uncertain. Some experts also contend that fossil groundwater would be depleted and that water supplies would therefore be seriously affected.

Arguments about water scarcity notwithstanding, cropping in these areas could be improved through scientific and efficient management of irrigation systems, the application of new irrigation technologies (such as drip irrigation), and through supplementary irrigation where rainfall is less than the total water requirements of the crop. Some of these new techniques have made irrigation feasible in areas where it was considered uneconomic only 10 to 15 years ago. Past experience in several African countries indicates that investment has so far

---

[31] Cameroon Rural Development Project Completion Report (Washington, D.C.: World Bank, 1986).

focused more on water engineering than on exploring alternative cropping systems and the associated problems of crop/water requirements in these difficult zones.[32]

Constraints To Irrigation in SSA

Irrigation development is essential for sustaining agriculture in SSA, but it is likely to be difficult and expensive to expand. Favorable conditions, such as good high-yielding aquifers, rivers with sustained year-round flows, and large tracts of irrigable lands are unfortunately not available to justify the type of massive investment that has gone into the Nile Basin, the Middle East, and Asia.

Physical. Irrigation projects in Sudan, Madagascar, Kenya, Ghana, and other SSA countries have already run up against physical impediments. The overall topography in much of western SSA is flat and offers few good dam sites, and all the rivers require storage work to meet dry season demands. This is the reason why most recent irrigation projects in SSA have been served by pumping from the rivers and not by simple gravity diversion. To be economically viable, such diversion structures would have to serve about 50,000 hectares, but no areas of this size and with suitable topography have been identified in this zone.

Another problem is that soils are often shallow and poor in nutrients. The large areas of residual soils found in the basement complex of SSA are suitable only for grazing, not for crop production. Where the soils are irrigable, large-scale schemes cannot be entertained because these soils tend to occur in small, scattered patches (except in Sudan). Furthermore, rainfall can be intensive for short periods, which means that surface drains in large-scale systems and spillways of dams must be designed with large capacities. Also, surface water resources seldom coincide with the areas in which irrigation could be advantageous. The ephemeral streams common in SSA provide little water for irrigated crops during the drought years as their limited flows must be devoted almost exclusively to supplying water for animals and humans. As for the possibility of using groundwater, this resource is localized, and there are no wide areas of shallow groundwater that could be exploited for conjunctive use.

Economic and Institutional. Potential investors in modern irrigation works in SSA are discouraged by the level of costs—which run to at least US$6,000 per hectare, and often much more, excluding roads and social infrastructure (see Tables 4.3 and 4.4). The high costs are related not only to the unfavorable physical conditions listed above, but also to economic and other factors. For example, most SSA countries have high taxes and artificially high values for their local currency. Added to this are high transport costs, which are caused by long distances and poor roads, fuel shortages, and inadequate transport services. Not surprisingly, foreign contractors (and the weak local construction industry in many SSA countries) often include in their bids a clause designed to protect them against the perceived risks of commercial operations in the region, particularly since government policies toward private investments are not always clear or guaranteed.

Contractors also complain that there are not enough local manufacturers, assemblers, dealers, or repair services for mechanical equipment. To make matters worse, in some countries the government plays a dominant but inefficient role in procuring and distributing equipment and machinery. Nor are there enough local managers and trained technicians. As a result, many systems operate inefficiently or must be supervised by expatriates, whose

---

[32] M. J. Blackie, *African Regional Symposium on Smallholder Irrigation* (Haerare, Zimbabwe: University of Zimbabwe, 1984).

charges push costs up even further. In addition, external financing agencies often insist that contractors comply with complex organizational and administrative procedures.

Table 4.3: ESTIMATED CONSTRUCTION COSTS FOR FULL CONTROL IRRIGATION IN FRANCOPHONE WEST AFRICA (1985)

(US$/ha)

TYPE OF SURFACE IRRIGATION	EARTHWORKS	CONCRETE	LAND LEVELLING	PUMPS	TOTAL
Village Schemes					
River terraces	1,000	700	-	1,500	3,200
Lowlands	2,600	1,800	400	1,000	5,800
Large Schemes					
River terraces	2,400	800	400	1,200	4,800
Lowlands	4,600	1,200	1,600	1,000	8,400

Source: FAO, 1986: Irrigation in Africa, South of the Sahara; FAO Investment Centre, Technical Paper No. 5: Food and Agricultural Organization of the United Nations, Rome p.42. Cost figures are based largely on Mali experience.

Table 4.4: TYPICAL RANGES OF CAPITAL COSTS IN NEW ASIAN IRRIGATION PROJECTS
(US$/ha)

COUNTRY	SMALL-SCALE COMMUNAL PROJECTS	MEDIUM AND LARGE PROJECTS
Indonesia	800	1,500-3,000
Korea	4,000-7,500	8,000-11,000
Nepal	-	2,000-6,600
Philippines	500	1,000-2,500
Thailand	50-500	1,500-3,000

Source: Leslie Small et al (1986), "Regional Study on Irrigation Service Fees", Sri Lanka: IIMI for the Asian Development Bank, p. 20.

Crop choice. The main crops produced in SSA under systematic irrigation are cotton, sorghum, rice, bananas, sugarcane, and vegetables; under flood irrigation, rice; and under receding floods, sorghum, mixed food legumes, and vegetables. However, the choice of high-value crops is limited. The principal cereal and legume crops are grown in the summer season. On the whole, wheat and other winter crops are not likely to be successful, except at high elevations, even when water is available. Thus, to improve crop intensity, farmers have to bank on short-maturing varieties that can produce more than one crop a year. Although rice and vegetable production has been intensified and improved in several West African countries, this has been done on a small scale. Tropical fruit production has also been advanced under irrigation. In Somalia, for example, banana production has been expanded in the lower Shebelli area for export to the Middle East and European markets, largely through private initiatives.

The production of groundnuts and cotton that grows under rainfed conditions has also expanded under irrigation in several SSA countries, but irrigation schemes incur such high costs that investment can only be justified if it ensures a substantial increase in the yields of traditional crops or of high-value crops. India, for example, uses 73 percent of its irrigation to produce wheat and rice—both of which carry a higher value than sorghum and millet. In addition, yields of irrigated sorghum are only 30 percent above the yields obtained from rainfed sorghum in most of SSA. This increase is hardly enough to justify the construction of irrigation projects for irrigated sorghum. Similarly, areas experiencing water shortages (e.g., California, Australia, Israel) tend to rely on irrigation to produce high-value crops and vegetables for export markets. Viable irrigation technology has been developed in water-scarce regions mainly for marketable crops grown under intensive cultivation.

In contrast to these trends, modern irrigation technology and systems of distributing water in SSA should be adapted to the needs of the crops already grown there. Yet, because suitable land is scarce and water expensive, planners find they must promote advanced technologies if irrigation projects are to be economically justified. Other commercial crops that could grow under the SSA agroclimatic conditions—such as sugar, cotton, and fruits—do not do well unless application is carefully managed, using a balanced mix of modern technology and flexible rotation of the traditional crops grown in SSA. For example, rice—the main irrigated crop in West Africa—does not demand precise irrigation (such as drip irrigation). Rice has been produced in several African countries (although at low yields) because it is suitable for traditional irrigation technology, which generally does not provide precise control over quantities of water to be applied in response to dynamic crop needs. Such technology can be extremely wasteful of water and may cause soil degradation as a result of water logging and salinization. In view of the increasing demand for more food in the region, the decline in soil fertility, and the limited supply of water, it is critical that traditional irrigation and water harvesting systems be modernized, in order to achieve greater economy and water-use efficiency and thus higher levels of sustainable production.

Many recent developments in irrigation technology could provide better economic and technical results. Some of these—for example, methods of applying water with frequency but in low volume, in precise response to changing crop needs—are already being employed in several parts of the world. Among these techniques are surge irrigation and moving systems of sprinkler, drip, and microspray irrigation. They seem to hold considerable promise with respect to increasing production efficiency. Properly applied, the new techniques could allow irrigation to expand into lands once considered submarginal (e.g., stony, sandy, sloping, or saline areas).

However, several criticisms have been leveled at these approaches. One complaint is that the new irrigation systems were developed in the industrialized countries, which are too capital-intensive and rely too much on high technology to provide appropriate techniques for the low-capital, low-technology, largely traditionally run production systems of SSA. Rather, the principles of modern irrigation must be adapted to the conditions of SSA. Moreover, the skills required to manage and operate these systems must be developed locally. The farmers themselves must learn to establish and operate these advanced irrigation techniques, but at present many of them have little or no idea of how such systems work, and they are reluctant to invest in such technology because they see that government irrigation agencies or dealers have had little experience in repairing and maintaining the new systems. These issues have been in the spotlight recently as irrigation planners and agronomists have come to realize that large-scale schemes may not be the right vehicle for agricultural development in SSA and that more effort needs to be put into installing efficient small systems.

Managing Common Water Resources For Irrigation And Water Supply

Irrigation is but one factor for planners to consider in the complex process of agricultural development. They must also take into account the user's right to land and water, the availability of production technology, and the type of system that will maintain, operate, and advance this effort. The most appropriate system will be one that is coordinated with the African tradition of common property resource management, and the institutions of this tradition. Some critics may argue that these institutions are moribund and should be replaced by more modern and efficient bodies entrusted with protecting the natural resources of the country. African tribes have been following their traditional system of land and water management for centuries. It is only recently, with the increasing population pressure and the rise of the central bureaucracy that these institutions have lost some of their influence. An important question for planners to consider is whether the traditional system of managing common property should be retained or replaced with a structure composed of modern technocrats and bureaucrats. The ideal lies somewhere between the two.

For generations, the typical system of managing a water point or flood basin in SSA has been to negotiate informal agreements among individuals in the same community or between two or more communities neighboring this resource. Land is usually allocated to farmers on a seasonal basis, and the user's right is determined by local leaders. In Sudan and Somalia, for example, land tenure is based on tribal boundaries. By tradition, each tribe has some jurisdiction over the use of land and water within its boundaries, which have been fixed in some cases through the village councils. Tribal units are divided into smaller units, each with its own territories.

One tribal group is usually dominant in each territory, but other groups also use the land there. Some of these groups utilize the land on a seasonal basis, whereas others have been using it steadily for generations, often under some land-borrowing arrangement with the dominant group. However, the burgeoning rural population is now struggling for greater control of SSA's resources and thus coming into conflict with traditional landowners. Another cause of these disputes in some areas is the expansion of modern agricultural schemes (irrigation, mechanized farming, ranching), which deprive traditional users of their rights. Since land rights derive from membership in a community, these conflicts at times arise between groups of people, such as pastoralists and cultivators.

Farmer Participation in Water Management

Several SSA countries have come to recognize the importance of having farmers directly involved in managing water resources, whether these resources are made available through the development of groundwater or the construction of small dams on seasonal streams. Where farmers have been allowed to manage their own water yards, operate them, recover the cost from the users, purchase fuel for pumps and spare parts for tanks and engines, the system has operated efficiently. Indeed, the people in SSA countries have operated and maintained wells for a number of years. Whenever the government has attempted to replace the traditional system and manage local water resources, it did not perform well because it had limited resources and a poor understanding of local cultures and traditions.

In addition, few countries have the institutions needed to manage their water resources properly. In other regions—for example, in Northern Africa where water is even scarcer than in SSA--thorough planning and targeted resource allocation have made it possible to develop several hundred thousand hectares into highly productive lands. This could also happen in SSA, with the help of both citizens and the governments of the various countries there. The role of the government should be to regulate, legislate, and rationalize the use of groundwater according to a plan supported by hydrological studies, land capability surveys, and land use patterns that will ensure a balanced opportunity for different enterprises, including livestock, forestry, crops, and grazing. However, such plans should not proceed with the support and participation of the users at almost every step in the long process of resource exploration, exploitation, management, and regeneration.

The experience with water users' associations in Asia could provide some useful lessons in this respect. Farmers in Asia have some control over their irrigation water, but the system could be improved if their organizations were more disciplined.[33] The power of any organization is usually a function of its ability to stimulate the participants and to channel their energies into group action, employing--with the agreement of participants--a reward and punishment system that reinforces the positive contribution of such energies. More field research is needed in Africa to determine how farmers could best be organized to manage water effectively.

Water development in SSA is currently hampered by the lack of information about local needs and resources. Up until recently, scholars and engineers were more concerned about agricultural development in Asia, where the Green Revolution, the advancement of irrigation, the evolution of farmers' groups, and the emergence of local institutions and other organizations have provided excellent opportunities for scientific investigation. The challenge for technicians and policymakers now lies in SSA, where much work needs to be done to improve resource management, local participation, and the balance between technical and social investment in agricultural development.

Although small-scale irrigation is largely a one-season activity, it would be beneficial to expand traditional and modern small-scale schemes in many parts of the region. Pilot units could be set up to provide farmers with credit or technical assistance to operate, maintain, and distribute water, and surveys could be made of existing water wells to assess their discharge and available groundwater, to study the communities using them, and to prepare models for expanding and enlarging these (where possible) on an experimental basis for crop, forestry, and livestock production. New and improved food crops (vegetables and legumes) could be tested under these pilot projects, along with organizational and financing models that would allow local users to manage and maintain their resources.

The Environmental Dimension

Irrigation development in SSA, as in any other region, has substantial environmental costs, primarily in the form of water-logging, salinization, damage to fisheries or water supply systems downstream. However, irrigation can cause two additional problems, both of which have been quite serious in Africa: water-borne and water-related diseases, and competition with floodplains.

The health problem. The incidence of disease has increased at a rapid rate since the expansion of irrigation in SSA. The most serious problem is schistosomiasis (bilharzia). Malaria is a distant second, but still a significant threat. Fortunately, other serious water-related

---

[33] Michael M. Cernea, <u>Putting People First</u> (Washington, D.C.: World Bank, 1985).

diseases, such as encephalitis, have not yet made their appearance in Africa, and on the positive side, the incidence of onchocerciasis (river blindness) has been reduced with the expansion of irrigation. The value of the healthy days of life lost (or at least the number of working days lost) due to the expansion of the diseases associated with irrigation should be quantified and added to the cost of projects.

Natural floodplains and swamps. Many river basins in Africa contain large floodplains that are important ecologically. Among these are the Sudd, on the White Nile (which is the largest swamp in the world), the Inner Delta of the Niger, Lake Chad on the Logone-Chari system, and the Okavango Delta in Botswana. These areas help to control floods, stabilize water flow, trap sediment, and are highly fertile areas for fish, meat, and grain production. In addition, large populations live and earn their living there, while many forms of wildlife, particularly migratory birds, live or sojourn there. These areas are also an important reservoir of vegetal diversity. Some experts argue, however, that the water in such areas could be put to other uses, such as expanding crop production and generating power.[34] Others believe that large-scale irrigation upstream of these areas would reduce the size of the swamps and thus the benefits they currently generate (e.g., some 100,000 tons of fish are caught annually in the Inner Delta of the Niger) and would jeopardize or destroy wildlife. It would also displace the large numbers of people who now live in and around the floodplains.

Obviously more precise information needs to be obtained on the environmental benefits associated with irrigation development to determine whether water diverted from swamps will indeed find more worthwhile uses elsewhere. Consider, for example, the results of a comparison of productivity per unit of water in the Inner Delta of the Niger and in the neighboring Office du Niger irrigated rice scheme. In terms of overall production, the former yields some 10,000 tons of meat, 120,000 tons of milk, 100,000 tons of fish, and 80,000 tons of rice, and the latter about 100,000 tons of rice. When computed per unit of water, however, irrigated rice produces almost 10 times as many calories and almost twice as many grams of protein as the swamp, although the swamp's produce is much more varied in type and nutritional value.

Recommendations For Developing Irrigation in SSA

It is imperative for the countries of SSA to step up their water conservation efforts and to find more efficient ways to develop their water resources. The measures that need to be taken to meet these objectives can be divided into seven broad categories.

(1) Improve the data base. Agricultural development depends on water. Development plans must therefore be based on a sound estimate of the country's water resources, both surface and groundwater, their suitability for agriculture and human consumption, and the feasibility of development and harvesting. Ideally, the plan should be formulated on a regional scale, but a country-level plan would probably be more practical. The World Bank, in collaboration with the United Nations Development Programme, has already initiated a hydrological assessment of Sub-Saharan Africa, but further support is needed to conduct detailed country-level studies of water and irrigation potential, and to support required action programs identified by these studies.

(2) Rehabilitate and modernize existing projects. One of the immediate tasks at hand is to protect the water resources already developed in SSA and to use them more efficiently.

---

[34] Jose Olivares, "The Role and Potential of Irrigation in the Agricultural Development of Sub-Saharan Africa" (Washington, D.C.: World Bank, 1987), draft.

Existing projects (at the country or regional level) should be assessed to determine how they could be improved and to propose the priorities for rehabilitation, expansion, or modernization. This effort should concentrate on getting the most benefit from existing investment and on dealing with technical requirements (spare parts, new equipment), policy issues (cost recovery, incentives, marketing, and pricing), and institutional questions (operation and maintenance, farmers' participation, extension and research in water management, and crop production). Thus the irrigation sector needs to be analyzed at the country level to determine what role it can play in stabilizing agricultural policies (food and cash), and in improving rainfed production through supplementary irrigation.

(3) **Conduct pilot projects to test new technologies.** The countries of SSA have had limited experience in supplementary irrigation. Therefore the first step they should take is to set up field-level models of water crop production and pilot schemes in groundwater development for crop, livestock, and human consumption. This would allow them to develop proper designs and to monitor and evaluate the pilot schemes over a period of, say, three to five years with the cooperation of local research centers supported by technical assistance, if necessary. The development of the Rahad irrigation scheme in Sudan provides a good model. Prior to large-scale investment, a pilot project was designed to assess different cropping and water management systems that could be incorporated in the final design of the Rahad scheme.

The highest priority should be given to expanding small-scale irrigation projects, especially in zones with poor potential for rainfed agriculture. Supplementary irrigation should also be investigated as a means of sustaining rainfed farming systems. In addition, large-scale irrigation should be carefully assessed as a possibility for selected locations, such as the Niger River. The countries of SSA need to acquire more experience on a small scale, not only with respect to the proposed technologies, but also with respect to the economic, social, and institutional aspects of implementation.

(4) **Improve cultivation practices as an integral component of irrigation technology.** Rainfed farming accounts for 80 percent or more of the gross agricultural product (GAP) across SSA. If supplementary irrigation technologies were adopted, the GAP and rural incomes would be increased by 15 to 20 percent, not to mention the multiplier effect that such increases could generate for the economy. The highest returns clearly come from supplementary irrigation, the next highest from rehabilitating the already installed infrastructure. New schemes should not even be considered until lower-cost technologies or production systems with higher returns have been identified. Furthermore, investments in human capital (individual or community) and in institutional development should be integrated in irrigation programs as a means of lowering their cost of water management, and of guaranteeing the viability of physical investments.

(5) **Put more emphasis on planning and monitoring water studies.** If water is to be considered a high-value resource, the planning of water usage (for human consumption, livestock, crops, fisheries, and urban industries) must be coordinated in the various ecological zones and monitoring systems should be established, especially for groundwater use. Attention should also be given to intercountry planning of water uses.

Water resources should be evaluated within a framework that distinguishes between marginal zones (or already degraded areas) and high-potential zones that require immediate attention and protection to prevent erosion and degradation. It will probably cost more to develop water resources in the marginal zones and the returns will be more limited. Investment in the high-potential zones could sustain and stabilize agricultural communities.

(6) <u>Conduct research on how to develop and improve irrigation.</u> Research should focus on the following areas:

- Technologies that could improve the efficiency of water management and the economic rate of return for irrigation. Water would thus become one of the main concerns of agricultural research.

- Crop varieties that require less water, and high-value crops that could have higher comparative advantage in SSA growing conditions.

- Market research, both domestic and international, with a view to diversifying products and gaining access to new technologies.

- Cultivation equipment to improve land preparation.

- Locally relevant organization and management practices.

(7) <u>Finance water management.</u> Water is considered a public asset in SSA. Although the users will have to bear operational costs--and possibly some of the development costs--there is little doubt that governments or communities and donors will have to bear most of the development costs. However, few SSA economies have the domestic capital to pursue such a course and need outside assistance.

(8) <u>Do not neglect the environmental and social factors.</u> The environmental and social aspects of water management also need to be studied in order to understand what impact the physical changes will have on the related environment and to determine what agriculture communities do and why they do it. Traditional custom or the lack of it should be investigated through field studies and household surveys as the resource base will be even more difficult to manage if the social factor is poorly understood, and careful research and scientific enquiries that strengthen the links between irrigation and the environmental sciences will help to avoid future disasters.

Many experts now agree that a concerted effort should be made to develop water resources in SSA, particularly in the arid regions. The problems already experienced with both large- and small-scale irrigation projects should not deter investors from supporting such projects. Rather, in-depth technical and social studies and fieldwork should be launched in every country of the region in order to identify relevant investment strategies. Modern field irrigation techniques have been spreading rapidly in the arid zones of the world, and the time is ripe to promote them in Sub-Saharan Africa, a region with escalating food problems that need immediate attention.

CHAPTER 5: SECTORAL STRATEGY FOR IRRIGATION DEVELOPMENT IN SUB-SAHARAN AFRICA: SOME LESSONS FROM EXPERIENCE

Uma Lele and Ashok Subramanian
(Uma Lele is Manager, Agricultural Policy, Africa Region Technical Department, and Ashok Subramanian is Consultant, Agriculture and Rural Development, World Bank)

Introduction

Irrigation is an extremely important potential source of stability and growth for agricultural production in Africa. Macroeconomic and broader sectoral policies should therefore provide the correct signals regarding the profitability of investment in irrigated areas. Such signals alone, however, are not sufficient to ensure efficient investments in a subsector like irrigation. Strategic decisions must also be made <u>within</u> the irrigation subsector if there is to be a proper fit between the irrigation program of a country on the one hand and the available resources and the strength of its institutions on the other. These decisions relate to the type and size of irrigation projects, mode of technology transfer, availability of technical and managerial capabilities, location of the scheme, project cost and financing, and project sustainability. Each of these factors needs to be considered in the context of a sound overall agricultural policy framework.

It is clear from the World Bank experience that many countries pursue more than one broad strategy in irrigation. For instance, Nigeria and Kenya have in the past adopted dual strategies--a primary program of large-scale irrigation projects and a secondary program of small-scale private and public initiatives. The question that arises in such cases is how the different strategies compare with reports of cost effectiveness and their contributions to agricultural growth and equity. The Bank and other donors have had enough experience in African agriculture to provide some valuable lessons that can steer future investments in the sector in the right direction. This discussion draws on the experience of Nigeria and Kenya in some detail, as well as that of Cameroon and Senegal. [35]

Many questions have been raised about large-scale irrigation strategies, especially about their economic efficiency, their impact on agricultural output and employment, technical viability, and the capacity of existing institutions to operate and maintain irrigation systems on a sustained basis where sunk costs through past investments make rehabilitation or additional investments tempting. In contrast, small-scale irrigation has had largely positive experiences and this offers considerable untapped potential for expansion involving both public and private investment. A number of measures can be adopted in support of private efforts by farmers. Where the public sector is involved, an effort should be made to decentralize irrigation systems and make their clients directly accountable for them. Operations will then become more manageable as users will have more control over important decisions concerning water availability and use.

As mentioned at the outset, an irrigation strategy will not automatically be successful just because of sound macroeconomic or even agricultural sector policies have been adopted.

---

[35] The review is based on <u>Managing Agricultural Development in Africa</u> (MADIA), a study supported by the World Bank and seven other donors. Ms. Uma Lele of the World Bank directed the study.

Explicit attention must be given to subsector strategy if the right decisions regarding technology, institutional development, and subsector investments are to be made. The "project approach" traditionally pursued by the World Bank and other donors can be a great help in the development of small-scale irrigation. Frequently, inadequate attention is paid to the irrigation subsector as a whole and hence to the need for a comprehensive approach to irrigation potential, or to the appropriate roles of large and small-scale irrigation and the steps required to develop this potential.

Rather, in countries where small-scale irrigation has been successful, such as Nigeria, this record of achievement has done little to shift the emphasis from the mainstream large-scale irrigation strategy in force. Powerful incentives exist to pursue the large-scale route: Political and economic power decision making, and the distribution of patronage that large-scale irrigation schemes entail are all centralized. Careful economic, technical, managerial, and political analysis should therefore be undertaken to test the relative effectiveness of the two strategies and the relative weights that should be assigned to different types of objectives in pursuing irrigation in the light of resource availability, alternative use of available funds, and institutional strengths of the country in question.

Irrigation In Nigeria And Kenya

The irrigation strategies of Nigeria and Kenya are particularly interesting because the macroeconomic and agricultural sector policy environments in the two countries are quite different yet their governments have both shown a similar fascination for large-scale irrigation.

<u>Macroeconomic and sectoral policies</u>. In the 1970s, Nigeria experienced a large increase in oil revenues that enabled it to significantly expand its public expenditures. The oil boom came on the heels of a civil war and displaced taxes on crop exports as a source of revenue. In the process, the federal government's revenues and expenditures increased sharply. Large urban expenditures drew labor out of agriculture. Overvalued exchange rates in the same period acted as a disincentive to agricultural exports. Government policy contributed to a sharp rise in food prices after the first oil boom and, despite rapid growth in food imports, food prices remained high relative to non-food prices throughout the 1970s and 1980s. Internal terms of trade between food and non-food crops shifted sharply in favor of food crops and this provided the impetus for increasing the size of the Nigerian government's investment in large-scale irrigation as a means of achieving national self-sufficiency in food and government revenues plummeted. Then in the early 1980s, oil prices fell, also reducing the resources available for large-scale irrigation. Consequently, a large gap developed between budgeted and actual public expenditures which reflected the lack of realism in planning investment programs out of the oil windfall so as to ensure maintenance and operations activities.

Not only was Nigeria's macroeconomic environment hostile to agriculture, but unfortunate policy circumstances also prevailed at the sectoral level. Political instability caused by a series of military coups and four years of civilian regimes contributed to numerous shifts in food policy initiatives. A government document acknowledged: "Past agricultural policies in Nigeria have been characterized by frequent changes or instability. This instability results not from some unpredictable exogenous factors, but more often than not, from changes in government or in the personalities of the operators of the system."

Kenya's overall macroeconomic policies have been more favorable to agriculture than those of Nigeria. Inflation has been more moderate and stable, exchange rate policies were generally satisfactory without serious overvaluation of the currency, and, Kenya's overall record of adjustment to external shocks has been more systematic and in the right direction. Nevertheless, the size of public expenditures and the governmental role in the economy have increased since the early 1970s, and have created domestic budgetary imbalances and uncertainties in the funding of projects, particularly the construction and maintenance of irrigation schemes in the agricultural portfolio.

Kenya has fared better than Nigeria in its agricultural sector policies. It was somewhat more consistent in emphasizing a smallholder strategy in agriculture. Land policies encouraged the registration and settlement of smallholders. Small farmers, who were denied the right to grow export crops during the colonial period, obtained those rights following Kenya's independence because grassroots interests had better representation in Kenyan politics. They were also able to earn international prices for tea and coffee, Kenya's major exports. The share of smallholders in coffee production increased from 35 percent in 1964 to approximately 60 percent in the 1980s (and the area they farmed incurs from 60 percent to 75 percent of total area under coffee). Similarly, smallholders in tea, who accounted for 5 percent of total production in the mid-1960s, contributed 48 percent of output in 1985. The experience with sugar, horticulture, maize, and many other agricultural products reflected a similar broadening of the participation of smallholders in production. Furthermore, Kenyan agricultural service institutions were relatively stable and more effective. The rapid growth in the adoption of hybrid maize in Kenya in the late 1960s represented a positive technological breakthrough.

In sum, on both the macroeconomic and sector policy fronts, Kenya has been relatively more consistent and effective in pursuing stable and equitable growth-oriented policies. Yet the Kenyan government followed the conventional route in irrigation.

Irrigation Strategy. Both Kenya and Nigeria have emphasized large-scale irrigation programs and have actively pursued them in the 1970s and 1980s. Nigeria's dualist irrigation strategy consisted of (1) a primary program of large-scale irrigation projects pursued under the direction of the River Basin Development Authorities, and (2) a secondary stream of small-scale irrigation schemes implemented through the <u>fadamas</u> and pump and tubewell operations in the Area Development Projects (ADPs). The "large-scale bias" in irrigation has meant that between 1973 and 1984, irrigation accounted for 40 percent of the government's agricultural expenditure. About 3 billion Naira were spent, but only 30,000 hectares of land (or 4 percent of total irrigated area) were irrigated—costing approximately $100,000 per hectare at the official predevaluation exchange rate. [36]

By the mid-1980s, small-scale schemes associated with the <u>fadamas</u> and low-lying areas accounted for about 94 percent of Nigeria's irrigation—covering about 780,000 hectares out of a total irrigated area of 830,000 hectares—most of it in the middle belt and the north. In contrast to the informal surface irrigation schemes, traditional water-lifting devices and ADP pumping schemes irrigate only about 10,000 out of a potential 930,000 hectares. The key governmental irrigation agency—the Department of Irrigation—has not been involved in these small-scale schemes.

---

[36] Lele, Uma, Ademola Oyejide, Vishva Bindlish and Balu Bamb, <u>Nigeria's Economic Development, Agriculture's Role and World Bank's Assistance, 1961-88: Lessons for the Future</u>, MADIA papers, June 1989. The precise costs of irrigation projects and the composition of these costs in terms of domestic and imported components are not well understood although it is clear that they have been higher relative to irrigation costs in Asia.

There is now increasing concern in Nigeria about the future of the large scale projects and their productivity. It has become evident that rates of return for small-scale irrigation could be two to three times that of large-scale operations, especially in the absence of downstream investments in canals and field channels to make water available to farmers. Indeed, in some cases, dams have affected the flow of water and water tables, and thereby surface and tubewell irrigation. Critics have questioned whether large schemes are indeed cost-effective and why it is necessary to rely on external expertise in the design and engineering of these projects, which means continuing dependence on external sources for technology and management. The government appears willing to review the large-scale irrigation investment portfolio and assess performance on a case-by-case basis.

Kenya has also promoted large-scale irrigation, but with very limited success, the worst case being the Bura irrigation project [37] which was developed between 1977 and 1986, planned as an integrated irrigation-cum-settlement scheme in a remote area with relatively low population densities, the project involved investments in several productive and social sectors. At the time of appraisal, irrigation was planned for 6,000 hectares and the World Bank was to meet about a third of the total estimated project cost of $98 million; the Kenyan government was to contribute about $21 million and the rest was to come from other donors. The project costs were high compared with the development costs of similar projects in Asia, even at the time it was planned. Bura was unsuccessful on many counts. The initial assessment of the irrigation potential was far too ambitious. Considerable technical problems were encountered. Significant cost escalations followed, and ultimately a much narrower segment of the farmer group was served than originally conceived. Between 1977 and 1984, the estimated costs increased by 30 percent from $98 million to $128 million; meanwhile, the area to be irrigated decreased from 6,450 to 3,900 hectares, with an actual per hectare cost of about $32,000 and the number of settlers in the project area reduced from 5,000 to 3,000. The final outcome in Bura was a negative rate of return. As the Project Completion Report for Bura notes "The Bura project did not contribute to the objective, expressed at appraisal, of developing Kenya's capacity to manage future major irrigation projects. On the contrary, the experience discredited development of irrigation as a tool to achieve agricultural growth and employment." [38]

There were many signs even at the appraisal stage that the Bura project should not have been pursued. Soil suitability was a continuing question throughout the course of appraisal. Salinity, high sodium content, and low subsoil permeability were recurring problems identified in project evaluation studies. Nevertheless, optimistic estimates of the potential outweighed the doubts and the project was launched, despite the warning sounded by Chambers and Moris in 1973: "There is a danger in the lower Tana of an irreversible commitment which given the heavy risks that normally attend organized irrigation, might .. [be] on a scale which would be a national disaster. In general, the larger the project is, the higher the cost, the larger the number of people involved, and the more publicity it receives. The risks are not simply that such a project will fail, but that having by all normal economic criteria failed, it will remain a permanent millstone weighing down the national economy."[39]

---

[37] Lele, U. and Richard Myers, <u>Agricultural Development and Foreign Assistance : A Review of the World Bank's Experience in Kenya, 1963 to 1986</u>, (Washington DC, World Bank) in press.

[38] World Bank Project Completion Report, <u>Kenya: Bura Irrigation Settlement Project</u>, June 22, 1988.

[39] Chambers, R. and J. Moris, eds., <u>Mwea : An Irrigated Settlement in Kenya</u>, Munich, 1973.

Although the rehabilitation of large-scale schemes often seems tempting, given the sunk costs, the costs and benefits must be thoroughly evaluated. Studies in Asia report that rehabilitation of old schemes may only mean further doses of ineffective investment.[40] Moreover, to ignore the development of institutional capacity before rehabilitation is undertaken only compounds past errors and leaves implementing agencies with a limited capacity to manage the complex restructuring processes involved in any rehabilitation effort. Rehabilitation, too, offers considerable rent-seeking opportunities and appeases political pressures. Its efficacy is difficult to evaluate objectively.

The complexity of designing a future strategy becomes clearer if one considers the location question. For instance, a study of population densities carried out for the MADIA project suggests that there is a need to carefully consider whether irrigation projects are being planned in areas of high or low population density; if in the low-density areas, it is important to ask whether investing in new projects or in the rehabilitation of old projects is really worthwhile in relation to the other alternatives available to address the employment and income-distribution objectives. Cost-effectiveness and the possibilities of labor migration to and from the project areas are two critical considerations. Locating projects in semi-arid areas in the Sahel where rainfall and population densities increases the costs of construction and maintenance. In addition, the transportation costs of moving output from these northern production areas to the major consuming centers in the port cities of the south can be very high. Even production costs tend to be higher on a unit basis in these areas. Although irrigation opens up the possibility of intensifying land use through the production of high value crops and multiple cropping, the rising demand for labor and labor scarcity in areas of low population density dictate the need for mechanization and centralization of a number of production operations. Under these conditions, the theoretical yield potential often cannot be achieved in practice without major subsidies. This lowers the returns on investment and substantially increases the budgetary burden, as in the case of the Flevue in Senegal or the Semry in Cameroon.

The Role of the World Bank in Irrigation. It is difficult to generalize about World Bank involvement in the irrigation sector in Africa except to say that its approval has been relatively cautious and has varied among countries, as decisions have changed over time, with the increasing recognition of the importance of irrigation. For instance, the Bank withdrew from Semry in Cameroon after financing two initial phases of the irrigation program because, despite high rice yields, production in Northern Cameroon for consumption in the port cities of the south was uneconomical, given the high costs of pumping water and transportation. Similarly, the Bank stayed out of the construction of dams for the Flevue in northern Senegal, but has recently been involved in helping to establish field channels in organizing rice production.

The Bank's approach to irrigation in Nigeria has also changed over time. In the late 1960s, the Bank was not well disposed toward irrigation in the belief that the Nigerian smallholder was yet to be oriented to the discipline of irrigated farming, economically and technically feasible farming systems were not yet available for irrigated farming, and the capital-intensive nature of earlier schemes had not provided attractive economic rates of return. In the early 1970s, however, Bank staff were impressed by the contribution of the traditional _fadama_ irrigation schemes that the Bank-funded ADPs promoted together with tubewell irrigation. A majority of economic benefits of the ADPs in the largely rainfed areas of

---

[40] Studies by Mark Rosegrant and others at IFPRI.

northern Nigeria now flow from small-scale irrigation that has promoted the production of rice and horticultural crops. Even in small-scale irrigation, however, much needs to be done to build the government's capacity to plan and implement small schemes.

The Bank stayed out of the activities of the River Development Authorities in Nigeria, which involve the interests of several states and require the direct involvement of the central government. The pressures for centralization are powerful in Nigeria--and elsewhere--and politically powerful domestic and foreign interests that benefit from the construction of large-scale irrigation tend to reinforce the centralizing tendency. Concerns for benefits in terms of agricultural production, cost savings, and employment tend to be displaced by rent- seeking and political objectives to appease powerful interests in the north. Despite the many demands for centralization and for the expansion of the large-scale irrigation portfolio, the Bank has been able to promote the sustained and consistent development of small-scale tubewell and surface irrigation through the help of excellent technical and managerial expertise.

The Bank's role in Kenya, in contrast, has not been too effective and, if anything, has been counterproductive. Despite the many doubts about the technical feasibility of the Bura project, it went along with the government, a path quite different from that taken in Nigeria. Having become involved in the venture, it has nevertheless remained an important participant in the project since it is difficult to withdraw abruptly from large-scale investment commitments with long gestation periods, even though changes in design and implementation have led to second thoughts about financing the project. In the final analysis, the Kenyan government and the Bank did not sufficiently recognize the lack of flexibility in undertaking such an investment; nor did they articulate the considerable risks associated with continuing and sustaining the project.

Lessons from Experience

The central lesson to draw from the African experience is that investment in irrigation cannot succeed without a carefully formulated subsector strategy. The strategy needs to include an assessment of soil and water potential, and of the technologies to be used for mobilizing and delivering water and in crop production; must take into account geographic and ecological attributes of other possibilities in locations of projects. Most important, it must insist on a review of the institutional capacity and human capital available in irrigation to handle investments and operations most efficiently.

The positive lesson of the discriminating role played by the Bank in Nigeria is that successful subsector work requires careful attention to issues of technology choice and modes of technology transfer. This involves a sound understanding of the resource endowments and strategic options facing governments. The Bura project in Kenya and the large-scale projects in Nigeria centered around expatriate consultants. The host countries did not have the expertise or the willingness to design and manage appropriate external technical assistance related to the dominant irrigation model, despite their strengths, as in the case of Kenya, in macroeconomic and sectoral areas. The Bank's staff associated with Nigerian small-scale irrigation seemed to have clearly understood the importance of a carefully chosen and consistent strategy for the subsector in the face of political pressures to the contrary.

Much greater priority needs to be assigned to building institutional capacity in analyzing, formulating, and managing both the technical and strategic components of subsector programs. Even in a more limited sense, a systematic assessment of the supply of irrigation engineers and systems and software specialists needed in Africa has to be made before major investments are undertaken. Rather than rely on inappropriate technology from the West, it is time to attempt innovative twinning arrangements with institutions in other developing countries, particularly in Asia and the Middle East, that have extensive experience in large, medium, and small-scale irrigation involving choices of technique more appropriate to the circumstances of Africa. A significant advantage here is that unit costs of irrigation projects can be reduced by the introduction of efficient indigenous technical and managerial manpower.

Although the Bank's strategy has worked and has proved successful in Nigeria, the dominant strategy of large-scale irrigation project development has been and continues to be central to Nigerian irrigation. Indeed, the mainstream irrigation bureaucracy under the River Basin Development Authorities--the prime mover behind the large-scale program--was able to insulate itself from the Area Development experiences and treat the small-scale irrigation of the FADAMA and ADP tubewells as a peripheral phenomenon. To further underline the power of this bureaucracy, the government has recently reaffirmed its commitment to retaining the organizational structures already in place even while questioning the viability of the large-scale projects. In Kenya, too, the expansion of the large-scale irrigation program continued in the midst of the increasingly problem-ridden Bura operations. The pace of expansion did not allow for the development of viable institutional capacity to deal with the technology transfer and sustainability issues inherent in such a program.

This raises the question of whether operating outside of the mainstream is likely to lead to substantive reforms in the overall subsector strategy in the long run. The limited experience of Nigerian irrigation suggests that a dual strategy can continue to function in a parallel fashion for some considerable time as long as there is no perceived threat to the mainstream model. It has taken a change of political leadership in Nigeria to finally review the relevance of the experience with large-scale irrigation. Raising questions about the dominant strategy is possible for donors but it is not possible to alter the strategy once the government and other donors ratify the investments. Donors, however, can play a catalytic role in initiating domestic debate and discussion on key subsector issues so that many views are aired before significant resource commitments are made. Technology choices associated with different scales of irrigation could certainly benefit from the internal debate on options and alternatives.

The redesigning of irrigation strategy in the 1990s will undoubtedly be a far more complex process than in earlier times. Any change in the balance of the strategy implies that the government's capacity to promote tubewell and small-scale surface irrigation must also be assessed and developed. A "promotional" instead of a "production" role calls for a major review of incentives within the public sector. Certainly, any analysis of alternative investments in irrigation must now explicitly consider location and the consequent effects of irrigated agriculture on the larger issues of growth and distribution; the costs of centralized control and indigenous institutional capacity for managing construction, operation and maintenance; and the environmental implications of various strategies. Any rethinking on strategy will clearly involve complex measures aimed at political and bureaucratic reorientation, away from politically and administratively attractive irrigation expansion paths to more economically efficient public and private investments.

PART II

THE LESSONS

CHAPTER 6: TECHNICAL ISSUES OF IRRIGATION DEVELOPMENT IN SUB-SAHARAN AFRICA

Akhtar Elahi and B. Khushalani
(Mr. Elahi is Senior Irrigation Engineer, and Ms. Khushalani is a Research Assistant - both from Africa Technical Division, World Bank)

Since 1980, donor support for irrigation projects in Africa has been declining. Among the primary reasons for this trend are the high capital costs of irrigation projects, their growing management problems, and the continuing controversy over cost-recovery policies. Despite these drawbacks, a cogent argument can be made for expanding irrigation in Africa, particularly in the Sub-Saharan region. This discussion is about several issues of concern to potential donors, some points they need to consider when evaluating proposed irrigation schemes, and three case studies of large-scale irrigation projects in Africa that offer some important lessons for future irrigation endeavors there. One of these projects was undertaken in Sudan and the other two in Madagascar.

Reasons For Choosing Irrigation

National planners are strongly attracted to irrigation as a means of supporting future food strategies wherever rainfall is marginal or erratic or where demographic pressure on rainfed land is rising. These conditions are all too common in countries of Sub-Saharan Africa (SSA), especially in Senegal, Mauritania (Senegal River Basin), Mali (Senegal River and Niger Basins), Ethiopia, Somalia, Sudan, Burkina Faso, Niger, Nigeria, and Chad. Irrigated agriculture can provide such countries numerous benefits: a greater degree of self-sufficiency in food and fiber crops, more equitable income distribution, a higher standard of living for their rural populations, a steady income for farmers, new employment opportunities, and reduced urbanization. Irrigated agriculture also provides farmers an opportunity to maximize production--through double or multiple cropping--and to take advantage of modern technologies and high-yielding crops that call for intensive farming under controlled conditions.

Perhaps the most important benefit in this part of the world is food security. The area that can be brought under irrigation may not appear large, but the contribution to food and fiber crops per unit area has been impressive. In SSA, the production value of an irrigated hectare is about 3.5 times that of a rainfed hectare. This improvement is possible because irrigation is an "intensive" form of agriculture and can produce surplus crops under uncertain climatic conditions.

The irrigation potential of SSA is estimated to be about 33 million hectares, of which amount 5 million, or 15 percent of the land, is currently under irrigation. About 2.6 million hectares are under large-scale modern schemes and 2.4 million hectares are under traditional small-scale schemes. More than 50 percent of the irrigated land is devoted to cereal production, 13 percent to fodder and 8 percent to fiber crops. Even though the area under production is only 6.5 percent of the total cultivated area, it is 20 percent in terms of total production value (Table 6.1). It is worth noting that half of the 5 million hectares are located in two countries, Sudan and Madagascar. The potential for a selection of SSA countries is shown in Table 6.2.

Table 6.1

TOTAL AREA UNDER RAINFED AND IRRIGATED AGRICULTURE AND PRODUCTION VALUES, SUB-SAHARAN AFRICA, 1980

	Area (M ha)	Percentage of cultivated area	Production value (M US$)	Percentage of value
Rainfed	116	93.5	29,376	80
Irrigated	8	6.5	7,475	20
Total	124	100.0	36,851	100

Source: "Irrigated Areas in Africa," Food and Agriculture Organization (Rome, 1987).

Table 6.2

IRRIGATED AREAS VS. POTENTIAL IN SOME COUNTRIES OF SUB-SAHARAN AFRICA

('000 ha)

Country	Irrigation Potential	Area Developed			Developed as % of Potential
		Modern	Traditional	Total	
Burkina Faso	350	9	20	29	8
Chad	1,200	10	40	50	4
Ethiopia	670	82	5	87	-
Madagascar	1,200	160	800	960	80
Mauritania	39	3	20	23	59
Niger	100	10	20	23	59
Nigeria	2,000	50	800	850	43
Senegal	180	30	70	100	56
Somalia	87	40	40	80	92
Sudan	3,300	1,700	50	1,750	53
Total	11,586	2,094	1,865	3,959	

The level of accomplishments to date can be seen in a scheme launched in Niger to restore the country's self-sufficiency in food after a severe drought in the 1970s. The government expanded the area under controlled irrigation, with rice as the main crop, and was able to increase yields considerably through double cropping in the entire project area. New technologies were introduced, two cooperatives were established, and research was undertaken on fertilizer use and pest control. The record of cost recovery has been good, and irrigated land was distributed equitably to 3,159 families, who were obtaining irrigated yields close to the appraisal estimate by the time of project completion. Similar results have been achieved in other large-scale projects, as shown in Table 6.3.

Table 6.3

EXPECTATIONS AND ACHIEVEMENTS OF SOME LARGE IRRIGATION PROJECTS IN SUB-SAHARAN AFRICA

	SEMRY I (Cameroon)	SEMRY II (Cameroon)	Lake Alaotra (Madagascar)	Morondava (Madagascar)	Rahad (Sudan)	Mopti (Mali)
Development area (ha)						
Expected	4,300	7,000	4,110	9,300	126,000	31,000
Actual	4,075	7,000	7,621	3,800	126,000	26,100
Cropping intensity (%)						
Expected	134	158	150	160	85	-
Actual	173	180	100	190	83	-
Incremental production (tons)a/						
Expected	11,460/4,500	47,000	25,900	32,500	119,000c/	35,000
Actual	18,335/13,500	53,000	24,500	15,800	114,000	29,400
Yields (kg/ha)b/						
Expected	3,000/3,000	4,000/4,700	3,350	4,000	2,020c/	1,860
Actual	4,500/4,500	4,500/5,500	3,000	2,600	2,278	926
Number of beneficiaries						
Expected	2,800	7,000	2,900	3,500	160,000	7,300
Actual	2,870	7,500	2,800	4,745	140,000	7,800
Economic rate of return (%)						
Expected	11	15	11	16	13.5	14
Actual	23	20	22	Neg.	20	17
Total cost/ha (US$)						
Expected	1,721	7,929	820	2,903	987	303
Actual	2,272	9,829	896	14,737	3,140	503
Total cost/family (US$)						
Expected	2,643	7,429	2,828	7,713	777	-
Actual	3,226	9,173	3,200	11,802	2,826	-
Implementation period (years)						
Expected	4.0	6.0	5.5	5.0	6.0	6.0
Actual	4.0	8.0	4.5	8.5	9.5	8.0

a/ All figure refer to paddy production, except where indicated.
b/ All figures refer to paddy production, except where indicated. Where two figures are given, these refer to wet season and dry season production, respectively.
c/ These figures are for cotton production.

What worries potential investors, however, is that these large projects have also experienced some problems, such as cost overruns on irrigation works, a lower than expected economic rate of return (ERR), and a slow uptake by farmers. These various problems should not be construed as a mark against irrigation itself. Like any development activity, irrigation projects are bound to suffer in an unhealthy economic environment, or in one in which institutional support is weak and inefficiency pervades both in the public and private sector. Some schemes in SSA that were productive at the outset have deteriorated for these very reasons. Table 6.3 provides information on the Bank's financial projects that were completed between 1970 to 1983. Of about 400,000 ha which have been brought under irrigation in these and other projects financed by the Bank since 1976, 300,000 ha are productive and managed by well-performing institutions (with ERR about 12 percent).

Another case in point is the 1.8 million hectares under irrigation in Sudan, which now require massive rehabilitation to bring productivity back to the level of the 1970s. The delivery system there has faltered in part because of inappropriate reservoir operations and the continued neglect of maintenance for the past 10 to 15 years. Pricing and marketing policies have done little to improve the situation, and in many cases the government has either been too closely involved in development activities, or not involved enough.

On closer inspection, it appears that almost all irrigation projects, like many projects involving significant amount of civil works, experience cost overruns, partly because costs are underestimated during the final design stage of the project. Overruns account in large part for the lower rate of return on many irrigation projects and explain why investors have become cautious and are beginning to overlook the potential benefits of these schemes. Donor confidence obviously needs to be restored--but the question is, how can this be done?

Points To Consider Before Investing In Irrigation Projects

Perhaps the decision to invest in irrigation in SSA has to be approached from an entirely new perspective--beginning with the realization that *capital works in SSA cannot be expected to* give their full returns soon after closing because of the time taken for full development under agriculture production. This point alone has large implications for the design of irrigation projects, but there are many others to consider as well.

Second, *planners must select an appropriate size of scheme for the given environment.* Large-scale schemes may seem to offer great rewards, but they may make little sense for small countries like Niger, which face formidable constraints to development, including poor soil and a wretched climate. Smaller schemes are more appropriate in such cases and should be undertaken at carefully chosen sites where less infrastructure and land leveling are required, and where participating farmers have a greater chance of recovering their capital investment.

Indeed, small-scale schemes (especially those operated privately) have had a better record of success than the large state-operated ones, which call for far larger capital outlays, a more complex infrastructure, greater expertise in irrigation engineering and technology, a larger distribution network, and more complex water management. That is to say, small-scale irrigation is easier to implement and does not depend for its management and operation on a central agency or large administrative structure that has a greater chance of running into problems.

Of course, small schemes have their own limitations, particularly in terms of area of land that can be developed and size of contribution to the economy. It is necessary to draw a judicious balance in the planning and development processes between small and medium/large irrigation schemes given the limitation of small schemes in harnessing the larger land and water resource potential offered in a specific area. Nevertheless, countries in which modern methods of irrigation are still in their early stages would do well to concentrate on small schemes to increase agricultural production. Such schemes can serve as pilot projects to provide much-needed experience in irrigation practices, agrotechniques, crop husbandry, adaptive research that can be replicated in the field, and in the types of crops that are most economic and most suitable for future large-scale development. Such projects also provide useful training for technical personnel and farmers, not to mention experience in water management, all of which will be essential for the success of future irrigation development.

Third, *success depends on an integrated approach*. No irrigation project, however well planned, can be expected to contribute to the national economy unless it has the support of the necessary satellite projects concerned with resettlement, the social infrastructure, roads, telecommunications, agricultural technology, research and extension, and government tariffs. These are usually set up as scattered subsidiary projects, whereas they should be incorporated into the irrigation project itself if maximum profits are to be achieved in the long run. However, donors should recognize that such a design is more costly at the outset because of its more numerous components. Moreover, such a design cannot work without the peripheral services of many organizations (public and private), ranging from fertilizer, insecticide, and pesticide procurement and distribution to agricultural credit facilities, access to markets, storage facilities, and pricing and marketing policies. All these activities have to be well-knit if irrigated agriculture is to be sustained. A number of irrigation projects in Africa have failed to pay sufficient attention to these details.

In turn, irrigation planning--which should include all the components outlined above-- must be integrated into the national water resources development plan. This plan must take into account all schemes--whether small or large, new or rehabilitated--and assign them their rightful priorities.

Fourth, *capital costs must be apportioned rationally*. As already mentioned, the physical conditions in Africa, the generally low level of development and lack of infrastructure keep the costs of irrigation development high in the countries of SSA. Consequently, extra effort must be put into the search for cost-effective designs, particularly for storage and diversion facilities, and for water conveyance, drainage, and distribution networks. In addition, planners should review case studies of earlier projects in the region to identify the causes of high costs elsewhere. In many cases, money is lost because methods are tried again, even though they have failed in earlier projects. Some planners have also made the mistake of departing from proven conventional methods for the sake of simplifying, as they think that this approach will help to reduce initial investment costs, but the results can be equally disastrous, particularly in the area of operating costs.

It is also important to note that the capital cost of an irrigation network will vary greatly, depending on the degree of controls embodied in the design. One must therefore identify the appropriate level of control for the given project. In addition, some capital costs should be seen as necessary development costs, particularly for facilities such as roads and communication networks.

Fifth, *cost recovery policies* have remained a source of continued controversy over the decades, largely due to the inability of governments and financing institutions to agree on apportionment of costs: social and equity elements on the one hand, and extent of recovery through direct and indirect charges on the other. Cost recovery policies should be conceived and accepted in advance. Maybe it would facilitate to delink cost recovery on investments and O&M charges on the project. O&M can further be eased by greater user participation and forming water users groups who should share the responsibility for regular O&M.

Sixth, *the technology must fit the African scene*. Countries have to find the right balance between a system that takes advantage of the state of the art in technology and one that takes into account conditions in SSA. The sophisticated technology required for drip and trickle irrigation, for example (used largely on vegetable and fruit crops), can be highly efficient, but the method cannot be used extensively in Africa.

Since the choice of technology has great bearing on the cost of construction, it should be investigated at an early stage in the project cycle. Planners should obtain informed opinions on the possible options from irrigation engineers, agronomists, and agriculturalists, particularly for projects requiring a sizable capital investment.

Here is a checklist of some other important questions that need to be considered before investing in irrigation projects:

- At the project preparation stage, have full investigations and surveys been conducted for topography, soil classification and suitable cropping patterns to be able to adopt appropriate irrigation technology?

- Are the designs detailed enough and not oversimplified for the sake of low initial investment?

- In new projects, has extra attention been given to developing basic management systems and to training of managers and staff?

- Is the government fully committed to the project?

- Has the appropriate degree of mechanization been examined and appropriately incorporated?

- Has the involvement and commitment of local and national authorities sensitive to sociological issues been sought?

- Does the project design make allowances for any changes which may become necessary at a later stage (e.g. size or scale)?

- Has the assurance of water availability been sufficiently examined?

- Will higher-value crops be added during crop rotation to increase cropping intensity and productivity to the extent allowed by marketability?

- Will time be devoted to adaptive research to develop economic crops for the particular ecological zone in question, and to develop suitable agro-techniques for field applications?

- Are the production costs compatible to make the project economically viable?

- Will an effort be made to recruit local technicians and extension agents who are familiar with local conditions and problems, and have good rapport with the populace?

- Have the land tenure aspects been examined to allow creating profitable productive units?

- Have appropriate alternatives been examined for a viable agriculture credit policy, including establishment of cooperatives?

- Have Water Users' Associations been established to ease the operation and reduce O&M burden on the public sector?

- Have analyses been conducted sensitive to social context to determine appropriate water users' fees?

- Will water pirating and water pilfering resulting in interfering with irrigation supplies and networks be stopped?

- Will there be an appropriate degree of private sector involvement to ease fiscal and management burdens to Government resources?

- Will cost recovery policies be formulated and accepted in advance?

A Review Of Three Irrigation Projects In SSA

(1) Sudan: Rahad Irrigation Project

The Rahad Irrigation project, approved in 1973, was the third such project to be supported by the World Bank in Sudan. The first large-scale project in Sudan was started in the 1920s, in Gezira. There are now five major irrigation schemes operating in the country: Gezira-Managil, New Halfa, Rahad, and the Blue and White Nile Pump schemes. The Rahad project provided for the development of the Rahad River, a tributary of the Blue Nile.

The project. The Rahad operation is located on the right bank of the Rahad River, some 150 kilometers downstream from its confluence with the Blue Nile. The project was the first phase of irrigation development in an area that became available for irrigation following the completion in 1966 of the Roseires Dam on the Blue Nile. Thus the principal source of water for the project is the Blue Nile, although seasonal flows from the Rahad are also used when available. Water from the Blue Nile is pumped from the main supply canal at Meina (see figure 6.1).

In addition, the project provided the social infrastructure needed to handle tenant families, housing for project staff, an internal road system, agricultural machinery and equipment, cotton ginneries, and a telecommunications system. The project was modeled along the lines of the long-established Gezira scheme. The construction and subsequent operation of the system was entrusted to the Ministry of Irrigation, while agricultural and other operations were taken over by a new project organization, the Rahad Corporation (RAHCO). Most of the irrigated land in the project area was allocated for cotton and groundnuts.

The Rahad project was originally proposed in 1963, but it took a long time to be approved. It was appraised in 1967, 1970, and 1972, and finally approved in 1973. It was turned down the first two times primarily because costs exceeded the funds available at the time. The approved project was expected to cost US$125 million. Soon after the project started, however, it became clear that actual costs would be far higher. Consequently, the project was reappraised in 1975, at which time it was estimated that total costs would be US$320.7 million, which was an increase of 156 percent over the 1973 appraisal.

This increase can be attributed to several factors. The project started up at the beginning of the energy crisis, which was also a time when inflation was running high. The cost of construction was also pushed up by about 30 percent because designs and quantities had to be changed since detailed design work had not been completed at the time of appraisal. Some of the increase was due to design improvements made during the appraisal. The donor agencies decided to finance the increase in project costs.

By the time the project was completed, costs had jumped to US$395.6 million, which was 23 percent higher than the 1975 appraisal. This increase was due mainly to the higher costs for building and irrigation works and upgraded specifications for project roads and telecommunications.

Outcome. Despite the cost increases, the final overall cost per hectare of irrigated land developed was only US$3,166--which, by current standards, is not unreasonable, especially for a project that includes extensive infrastructure. In addition, the project has succeeded in developing the entire area of irrigated land identified at appraisal, and, in general, the scheme has been completed satisfactorily. However, because of the long delays during implementation, the project was completed about four years after the originally specified date. Since then, some technical problems have arisen at the main pumping station, and it has been necessary to repair cracks in the lining of the discharge basin and to deal with recurring siltation in the intake channel. The settling of the foundations has also caused some cracking in the project buildings.

During project appraisal, the ERR was calculated at 13.5 percent, but this figure was raised to 16.5 percent during the reappraisal, and subsequently to 20 percent because of the good agricultural performance of the project. However, substantial amounts have been invested in Roseires Dam, which had to be built before the project could begin. The economic analysis for the project therefore treated the cost of building the dam as a sunk cost. Although this approach is correct, the ERR from Rahad would have been lower if the cost of the dam had been factored into the project.

Production under the Rahad project has been organized around the pattern at Gezira. Under this arrangement, the government is responsible for developing and providing irrigation, a parastatal corporation oversees management services, and tenants supply labor and perform certain field tasks. This system takes into account the limited experience that Sudan farmers have with irrigated agriculture.

During the 1970s, about 50 percent of gross agricultural export earnings were derived from cotton, mostly from the irrigation schemes. Groundnuts, sesame, and gum arabic contribute 25 percent. Foodgrain and livestock exports from the rainfed sector--which have increased sharply in recent years--accounted for 19 percent of the volume of merchandise exports in 1979-80. The most striking feature of agricultural exports from 1975 to 1980, however, has been the sharp decline in the volume of traditional cash crops, cotton, sesame, and groundnuts. This reduction has contributed greatly to Sudan's current economic difficulties. The decline occurred in part because management did not pay enough attention to maintenance. Crop diversification and intensification in the 1970s also put a strain on the delivery of inputs. Meanwhile, skilled manpower was fleeing to the neighboring oil-producing countries. At the same time, the government was keeping the producer price for cotton low in order to maximize short-term budgetary revenues.

To correct the growing imbalance in the export account, the government in 1979 introduced some important policy reforms, particularly with respect to cotton pricing and producer incentives, in order to stimulate production. A rehabilitation strategy was designed to concentrate investment efforts over the coming years in the irrigated subsector in order to arrest the decapitalization that had set in during the 1970s. To date the effort appears to have been successful—cotton production in 1981-82 was 59 percent higher than in 1980-81, and production in 1982-83 was 28 percent higher than in 1981-82. The latter compares with the average annual production of 628,000 tons during 1970-75. The Rahad project contributed about 21 percent to Sudan's cotton production in 1982-83.

The World Bank's role. Credits for agriculture form a large proportion of total World Bank lending to Sudan, averaging half the lending through the 1970s up to the present. This support has covered six irrigation projects, three rainfed mechanized farming projects, three smallholder development projects, a livestock marketing project, and agricultural research project, and an agricultural rehabilitation program. The World Bank also helped finance the Roseires Dam, completed in 1966. The ERR for that project was 14 percent, and its net socioeconomic impact—for the country as a whole and for most of the population directly affected—was positive. However, the investment in Roseires Dam was thought to be about five years premature, in that downstream investment plans were not detailed enough at the time of appraisal. The World Bank is also financing several rehabilitation projects in Sudan.

Lessons to be learned. Projects in the agricultural sector of Sudan, as in most other sectors, have been difficult to implement. Basic materials such as cement, fuel, and imported farm inputs have not been available when needed because of foreign exchange shortages and transport problems. Disbursements have been slower than forecast, and consequently closing dates have frequently been extended. In addition, gestation periods have been longer than expected, with a resulting lag in the arrival of benefits.

Although the ex-post ERR for Rahad has been a respectable 20.4 percent, it could have been still higher if the implementation delays had been avoided and the appropriate roads and telecommunication components provided on time. Although later delays were due mainly to local cost financing problems, the earlier ones were the result of inadequate project preparation, unrealistic estimates of costs and quantities during project appraisal, and poor organization of implementation measures at the start of the project (largely because the government's ability in this regard was overrated). Worldwide inflation and the oil price shocks of the 1970s also contributed greatly to the cost overruns, but these events could not have been predicted at appraisal. However, other constraints were recognized during the 1975 review and steps recommended to rectify them. Although the end result was satisfactory, the 1975 review did not adequately address the problems surrounding roads, telecommunication, and building components, which therefore have not yet been resolved.

One vital lesson to be learned from this experience is that the execution of large and complex projects must be carefully planned between the feasibility and implementation stage. Thus, since 1981 the World Bank has required field investigations and detailed designs to be well advanced at the time of loan approval for all types of projects. The second lesson is that very large irrigation projects requiring settlement, a complex infrastructure, a new agricultural organization, and capital works cannot be completed in Sub-Saharan Africa within the normal loan period of five years.

(2) Madagascar: Morondava Irrigation and Rural Development Project

The Morondava Irrigation and Rural Development Project was approved in June 1972 and was the World Bank's third agricultural lending operation in Madagascar and the second for irrigation development. The project was implemented over a 10-year period, which witnessed many institutional and policy changes that were to have a wide-ranging impact on project performance and on agricultural production throughout the country.

The project. The project provided for the rehabilitation of an existing irrigation network in the Morondava River Basin and agricultural production activities in three subunits of the irrigation scheme. It was to be the first phase of a long-term development program aimed at irrigating 30,000 hectares on the southwest coast of Madagascar. This area was thought to have great agricultural potential if water availability could be regulated, and to be suitable for resettlement because of its low population density.

The primary project objectives were to diversify the production of export crops by introducing tobacco and groundnut cultivation and to increased paddy production (from 1,800 to 30,400 tons per year) and cotton production (from 750 to 4,200 tons per year). The overall responsibility for project implementation was assigned to the Société pour le développement économique de la région de Morondava, a parastatal company created for the purpose. The Rural Engineering Department of the Ministry of Rural Development was in charge of major civil works and the construction of the drainage system.

The project was part of the government's program to accelerate regional development and to encourage settlement of the underpopulated but fertile areas of the country's west coast. It also supported the government's goals in the agricultural sector--namely, to increase rice production (the staple food crop) and to diversify export crops. The project was part of the government's overall drive to rehabilitate and expand cultivated areas under large-scale, modern irrigation schemes. However, the new governments installed in 1972 and 1975 questioned the appropriateness of these goals and instead put more emphasis on helping the small farmer and on supporting small-scale irrigation schemes.

The central element of the project was the construction of an expensive diversion dam, which was considered essential if irrigation activities were to be expanded in the Morondava River Valley. By way of support, primary and secondary canals and drains were to be rehabilitated, rural roads constructed, and on-farm development works initiated. In addition, two state farms were to be created for cotton, tobacco, and groundnut cultivation. Project investments were phased over a five-year period and full development was expected to be achieved in year 11 after the start-up of the project.

During preparation, however, the project area was expanded from 5,00 to 10,000 hectares, and high-value cash crops were introduced. This appeared to be the minimum package that could justify the high cost of the proposed civil works. Detailed engineering studies for the major civil works were also begun, and by early 1973, it was clear that the design and phasing proposed in the appraisal report had some serious problems. For one thing, the location of the canal intake was unsuitable, and the existing settling basin would be unable to prevent the canal network from sanding up. For another, the gradual phasing of investments proposed earlier was not feasible. Consultants recommended that the design be modified in order to correct these deficiencies, but these changes (combined with the devaluation of the U.S. dollar) were expected to push the costs up to 100 percent of the initial appraisal estimates.

After the credit was declared effective, the civil works continued to be the focus of attention, although research, cotton production, and settlement activities had also started by this time. A hydraulic model of the design for the civil works component was tested, bid documents were drafted and finalized, and bid invitations were advertised in October 1973. However, the bids surpassed the revised cost estimates (even the lowest bidder was 15 percent higher). The government entered into negotiations with the lowest bidder and in May

1974 a contract was signed. When the contractor drilled boreholes and conducted permeability tests along the weir site, it was discovered that the construction techniques proposed could not be used, and that different (and more expensive) techniques were required. The contract was subsequently renegotiated, and the final price (excluding taxes and provisions for price escalation) totaled US$15.5 million, or 140 percent more than the initial appraisal estimates.

These cost increases alarmed officials, who saw that the government would have to finance a much larger share of total project costs, as well as World Bank staff, who began to question the economic viability of the scheme. When in August 1974 the government requested a supplementary credit to finance cost overruns, World Bank management decided to send a supervisory mission to evaluate not only the request, but also to reappraise the entire project.

Revisions to the project. The supervisory mission examined all aspects of project implementation--including the civil works, settlement activities, agricultural production, implementation techniques, and economic justification. The mission found that the costs of many other components besides the civil works had increased substantially. In fact, the total costs of the project had doubled. Meanwhile, benefits had not increased commensurately, and it was doubtful whether certain components (land development/consolidation, tobacco farms) could realistically be implemented in the time frame proposed. The mission concluded that the economic returns would probably be negative, or at best marginal, and that implementation as originally designed would represent a serious misallocation of the government's resources. The mission therefore recommended that the government cancel the civil works contract (the order had not yet been given), scrap the existing project, and begin preparing an entirely new project based on a different approach to the development of the Morondava region, and possibly forgo further irrigation development in favor of rainfed food crops/rural development.

The World Bank denied the government's request for supplementary funding and attempted to persuade authorities not to continue with the project as originally designed. However, political disturbances starting in February 1975 interfered with communications, sometimes for weeks. During this period, senior Bank management reviewed the project's problems and agreed with the members of the mission that the civil works contract should be canceled. However, they found that there were no legal grounds for suspending disbursements on the contract.

A high-level delegation was sent to Madagascar in October 1975 to assess the situation and to devise a plan of action for project implementation with government authorities. The mission found that civil works construction had begun in March 1975 (immediately after the end of the rainy season), and that halting project implementation would result in a loss of more than US$7.0 million. Thus, halting implementation was no longer an option, and the second best solution appeared to be to redesign the agricultural development components in order to reduce costs and the financing gap. The mission was not willing to recommend additional financing from the Bank to cover cost overruns, as mission members continued to believe that the project was not economically justified and should not have been implemented. Thus the two parties agreed to a package of civil works and supporting activities that would permit development to continue within the confines of existing financing and with relatively small increases in the government's contribution.

The redesigned project was expected to cost US$56 million, which represented an increase of US$3.7 million to be borne by the government. The ERR was estimated at 1 to 3 percent, owing to the high cost of the civil works and the relatively limited benefits. The disbursement schedule was modified to reflect the changes in the project design and to ensure adequate financing for all components.

Outcome. Construction of the main civil works was completed in February 1980, but performance in other areas continued to be disappointing. The implementation of on-farm works fell behind the pace needed to meet the scheduled deadline, and the construction of buildings had virtually come to a halt. Despite the improvements in farm management, agricultural production also turned out to be disappointing, owing to bad weather and lower producer returns. However, research activities were well managed, and a program was initiated to identify relevant and cost-effective technical packages for extension to area farmers working with irrigated rice. In an attempt to strengthen research and extension links, the government also established demonstration plots with local farms in the newly developed areas. But other support services ran into problems: land preparation services had to be halted and fertilizer and tool sales fell because credit was cut off. The irrigation network was virtually unmanaged as water pirating had become commonplace and local farmers were beginning to ignore the recommended cropping calendar. Maintenance also suffered owing to the lack of equipment and operating funds.

The agricultural impact of the project has been less than anticipated at appraisal, and has been mixed when compared with reappraisal estimates. Farmers have made only a limited effort to adopt improved cultural techniques based on research recommendations, and despite the sophistication of the irrigation works, water management has fallen far below the standards required of a modern irrigation scheme. Nevertheless, the project has helped to increase production as well as farm family benefits.

Lessons to be learned. Water management and network maintenance are vital to the success of an irrigation project. These operations must improve in the Morondava case if production is to increase. Water use planning will also have to be introduced as competing demands for network water will soon exceed network capacity.

(3) Madagascar: Lake Alaotra Irrigation Project

This is the first time that a performance audit has found a project that was classified as "worthwhile" after completion to be a failure some years later, and to report a negative re-estimated economic rate of return.

The project. In June 1970, Madagascar received a US$5 million credit to rehabilitate and expand the irrigation and drainage network in an area of the Lake Alaotra Basin. The project was designed for an arid area where the average annual rainfall is 1,000 millimeters, most of it occurring between October and April. The overall goal was to rehabilitate existing canals and drainage networks on 4,110 hectares of already irrigated rice fields. There was to be subsequent clearing, drainage, and irrigation, and a road network was to be constructed on an additional 5,910 hectares of "unimproved" land, some of which was already being used for non-intensive rice cultivation. In addition, about 2,000 hectares (some also being cultivated non-intensively) would be cleared and drained for dry-farming of rice. Three other components of the scheme were on-farm development, a rural development study, and crop diversification.

The project had two primary objectives: (1) to increase rice production by improving yields and extending cultivable and irrigated surfaces, and (2) to diversify agricultural production in a region (erroneously) seen to be an exclusively rice-growing district. These goals were to be accomplished by double-cropping on 5,000 hectares of the affected rice fields. Farmers were expected to triple their incomes, unemployment was to be reduced, secondary economic activities increased, and an efficient extension service started.

Project costs were estimated at US$8.2 million (which increased to US$8.9 million when variations in the exchange rate were taken into account). The World Bank group was to finance 60 percent, and the government and the Banque Nationale Malagasy de Développement were to provide the remaining funds. The Ministry of Agriculture and the Société Malgache d'Aménagement du Lac Alaotra (SOMALAC) were responsible for implementing the project.

The major civil works were completed on time, but on-farm development, which was to be carried out on force account by SOMALAC, was never finished. The final area added by the project was 3,511 hectares (rather than 5,910) of irrigated land, and 1,793 hectares of land cleared and drained for dry-farming. The current irrigated area is 7,621 hectares, or about 75 percent of the intended area. The 1,793 hectares "improved" for dry-farming is close to the project goal of 2,000 hectares, but much of this land has been destroyed by excessive drainage because of an accident that occurred during project construction. Consequently, the outcome has been just the opposite of the expected increase in productivity.

Project managers recognized that yields and total production were below appraisal estimates. However, the higher world market price of rice between 1973 and 1975 and the belief that the full projected area (12,000 hectares) had been developed and cultivated, led managers to think that the ERR would be about 10 to 30 percent, or about twice the appraisal estimate. Poor cost recovery of planned irrigation needs and land recovery charges had created severe financial problems for SOMALAC, which had been entrusted with distributing land, maintenance, and extension. By 1981 it was obvious that few if any of the project objectives would be realized. The actual ERR was probably going to be negative, and incremental annual paddy production over the without-project estimates was only about 2,000 tons, rather than the 26,000 tons forecast at appraisal.

Outcome. The project has made no significant contribution to the Malagasy economy. Because of population growth and immigration, more rice is being used to feed local people than is being exported. There has been no crop diversification or double-cropping. Farm incomes have barely increased, and wealth differentials are much greater than when the project began. There has been an increase, rather than a decline in joblessness; secondary activities have not increased; and there has been little effective institution building.

Cost recovery has been very poor, and SOMALAC continues to experience severe financial problems, which have limited its effectiveness. Nevertheless, there is some evidence of increasing intensity of rice production in the project area. However, the reasons for this were unforeseen at appraisal. Immigration and local population growth are putting increasing pressure on the project to provide subsistence as well as cash incomes. In contrast to many other development projects that have promoted a shift from subsistence to cash economies, intensive subsistence farming is increasing here, although commercial production is still substantial. This intensification may eventually lead—by some unknown route—to the project goal of increased rice production. However, there was no evidence that it had done so by 1981.

Lessons to be learned. The current low productivity is related to the poor maintenance of the irrigation network, the lack of inputs, ineffectiveness of the extension service and of SOMALAC in general, and the low rainfall. However, the main reasons for the failure of the project appear to be technical, environmental, social, and demographic. To begin with, the irrigation canals were dug lower than the rice fields they were intended to irrigate, and are sinking further into the peat soils. Second, there has been too much drainage on the previously marshy lands, which are now classified as dry-farming (pluvial) areas. Farmers have attempted to control the weed growth attributed to this excessive drainage by their customary practice of burning at the start of the rainy season. The peat soils, rendered drier as a result of mismanagement of the project, have caught fire, with the result that productivity is declining. Third, the area's hydrology has probably changed, and siltation has increased in the irrigation system because of increased erosion in the hills west of the project.

Perhaps the most surprising outcome is that demographic and social factors, scarcely mentioned at appraisal and still poorly understood at completion, have marred the project's chances of success. At the outset of the project, the area had a sparse population (about 15 per square kilometer), and it became necessary to bring in both seasonal and permanent workers to provide labor for the rice cultivation. This change has created a diverse population, but also conflict because of ethnic and socioeconomic stratification. Now, five years after the completion of the project, the growing population is promoting intensified land use. Immigrants and locals who had no land are farming smaller and smaller plots through various informal, and sometimes illegal arrangements (renting, sub-renting, or sharecropping). As the population has grown, the tonnage kept aside by the farmers for their subsistence has increased. And since total production has not increased very much, if at all, the marketed quantity has decreased. In sum, the project has benefited immigrants more than the Sihanaka natives, whose traditional generalized economy (based on fishing, gathering, herding, and non-intensive cultivation) is being destroyed because of the project.

Conclusion

Past experience with irrigation projects in Africa suggests that such schemes can produce many economic benefits. The problem is that a great deal more attention needs to be given to the other aspects of irrigation projects besides construction. Otherwise the benefits will not be realized. The irrigation plan itself, for example, must be conceived in a long-term time frame. This will make it possible to take into account and plan for local constraints on implementation, and also develop a "pipeline" of quality projects for the future. As many past projects have demonstrated, maintenance and operation cannot be neglected. And social and demographic problems should receive just as much attention as the technical ones.

This is not to say that technology should receive less attention. In fact, it is extremely important to choose the appropriate technology for the SSA setting, whether the system being considered is a new or a refurbished one. Time must also be spent studying case histories of existing projects, particularly those operating at less than desired levels, in order to determine why their performance is poor and to make certain that similar mistakes are not repeated. Perhaps local managers need more training, or the agricultural and extension services are weak, or support services deficient. That such problems have occurred should not deter potential donors. The important point is that they be identified and brought to everyone's attention during project appraisal so that they can be avoided in the future. These and other problems can be overcome, and sustainable irrigation can be developed in SSA, but the task is a complicated one that cannot succeed unless the countries of the region demonstrate to donors that they are committed to their development objectives and are willing to take an integrated approach to the management of their water resources.

CHAPTER 7: POLICY AND MANAGEMENT PROBLEMS

Shawki Barghouti and Ashok Subramanian
(Mr. Barghouti is Division Chief, Agriculture Production and Services and
Mr. Subramanian is a consultant both working in the Agriculture and Rural Development Department, World Bank)

The experience of irrigation projects in Africa has clearly demonstrated that such projects cannot thrive in an unfavorable macroeconomic and sectoral policy environment. Overvalued exchange rates, suppressed producer prices, deficit-induced reductions in capital and operating budgets of public programs and other similar governmental interventions have stood in the way of success in many irrigation projects. Unfavorable policies have stalled important stages of construction, hampered critical maintenance operations, and prevented producers from making the right cropping and marketing decisions, and thus have reduced their potential incomes.

There is little doubt that corrective policy action on the part of governments is a desirable step for African irrigation. But even as we address important problems of policy, we have to pay continuing and perhaps even increasing attention to irrigation project development and management, which are also major determinants of project success or failure. The need for such attention can best be demonstrated by the experiences of irrigation projects in Sub-Saharan Africa, as documented in World Bank Project Completion Reports (PCRs), Project Performance Audit Reports (PPARs), Impact Evaluation Reports (IERs), and Sector Review Reports (see Table 7.1).

Table 7.1

Estimated ERR of Some Irrigation Projects in Sub-Saharan Africa

COUNTRY	IRRIGATION PROJECT	ERR (%)
Madagascar	Lake Alaotra	Negative
Sudan	Rahad	20.0
Mauritania	Gorgol Noir	2.7
Kenya	Bura (settlement)	-13.0
Niger	Niger	3.0
Burkina Faso	Niena Dionkele (rice)	-5.3
Cameroon	Second Semry (rice)	16.0
Senegal	Debi-Lampsar	-4.0
Senegal	Polders	--
Sudan	Roseires	14.0

Policy Environment

Past experience in Africa clearly demonstrates that an unsupportive policy environment greatly reduces the chances of project effectiveness. Early in the Sudan Rahad project, cotton prices were poor owing to problems with the foreign exchange rate applicable to cotton transactions. The government pricing policy left an insignificant balance for distribution to farmers after all the deductions had been made. The net amount paid to farmers for their cotton was very unattractive and farmers showed little interest in cotton production. As a result, marketed output was lower than expected. This matter was settled in 1979 in part by restoring the foreign exchange rate with respect to cotton to realistic levels. Subsequently, the marketed output increased.[41]

The important role of price incentives in agriculture can also be seen in Niger and Somalia. In the latter case, the government was torn between the incentive needs of producers and the affordability needs of urban consumers.[42] In 1981, the producer prices for cotton, sesame, and bananas in Somalia were far below world market prices, which had a negative effect on irrigated crop production.

Until the early 1980s, the public sector in Somalia had a large and critical role to play in several agricultural enterprises. Its performance with regard to the public enterprises was by and large unsatisfactory. Land tenure and registration policies and procedures were deficient in many respects, as were output pricing by marketing enterprises, input distribution systems, and agricultural research.[43]

In Senegal, the rising price of rice and groundnuts helped output growth. But the rising prices of inputs did not. Burkina Faso had to contend with inadequate long-term central government support for a project that produced crops directly competing with cheaper imports. In the early 1980s, domestically produced rice was selling at 130 CFAF per kilogram, whereas imported rice cost only 100 CFAF per kilogram. The mill that was procuring rice from the producers needed a great deal of government support, but only a limited amount was made available owing to fiscal constraints.

Fiscal Policy

Of more direct and immediate significance to irrigation in a managerial sense is the effect of a government's fiscal policies on irrigation operations. Budget constraints in several countries have caused a severe strain on project funding. Although budgetary decreases affect all sectors of the economy, they tend to hit irrigation particularly hard because of its large capital and operating needs.

Budgetary constraints forced authorities to reduce the scope of an irrigation project in Kenya, and the lack of counterpart funds was responsible for the poor maintenance and deteriorating perimeters of the Senegal River Polders Project. In the Sudan Rahad project, the government did not provide adequate support for operation and maintenance in the early years. Consequently, silt accumulated in the canals, and eventually the Sudanese government was forced to maintain the drainage system properly.

Indeed, better cost recovery procedures and effective implementation could have offset the effect of budgetary reductions in all these irrigation projects. Yet the cost recovery program was quite inadequate in Rahad, as in other early projects. The lessons of

---

[41] OED, Project Performance Audit Report 5130, Sudan: Rahad Irrigation Project, June 13, 1984.

[42] Somalia Agriculture Sector Review, Report 2881a-SO, Volume I, June 29, 1981, p. 17.

[43] Agriculture Operations Division, Eastern Africa Department, Somalia Agricultural Sector Survey, Main Report and Strategy, Report 6131-SO, December 30, 1987.

these projects were incorporated into subsequent efforts. As the PPAR for Rahad observed: "The Bank's experience in Sudan subsequent to the Rahad project has been to negotiate the principles rather than the rate of return and the levels of land and water charges. Thus, covenants to later projects have indicated the sort of costs to be recovered and the extent to which they should be charged to farmers. Further, events such as rapid inflation have made annual adjustments to land and water charges imperative."[44]

The World Bank's experience with irrigation operations indicates that some fundamental covenants on cost recovery have by and large been ignored. Many reasons have been given for this lapse. As pointed out in a recent OED review, "Adequate O & M of irrigation systems is a prerequisite to productive farming which in turn is a precondition of cost recovery."[45] Thus water availability is an important precondition to effective cost recovery. At the same time, there is a need for (1) government, donor, and farmer commitment to project sustainability and hence to cost recovery, and (2) cost recovery policies in the context of the other taxes and subsidies operating in the agricultural sector.

In short, cost recovery policies and procedures require the utmost attention in projects. But putting them together is not a simple and straightforward matter, especially in the sectoral context. What is clear is that a _necessary condition_ for the implementation of effective cost recovery is _good project design_ that ensures water availability and _good institutional development_. This is the only way to ensure that farmers will have access to water resources and to assure sustainable programs of irrigation.

Project Management and Performance

Poor performance on many policy-related variables—exchange rate, producer price, other related prices, extent of use of the public sector, fiscal policy—has contributed to an unfavorable economic rate of return (ERR) in projects, but so have project management and implementation issues. It is, of course, difficult to specify what particular variable or set of factors is directly responsible for the ERR of a project. By and large, the negative or very low ERR estimated after the close of the project either at the time of the PCR, PPAR, or IER has been due to cost overruns, lower-than-expected volume and reach of irrigation and agricultural services, and poor farmer response to recommended income-increasing practices. The most dismal performance—in the Kenya project, which had the lowest estimated ERR—was the result of many factors, topped by poor project design and implementation.

Of the projects reviewed, the following had the lowest estimated ERR (all negative): Kenya Bura Irrigation Settlement, Madagascar Lake Alaotra Irrigation, and Burkina Faso Niena Kionkele Rice Development. Although policy flaws were partly responsible for this outcome, the principal reason for the low ERRs was inadequate project design and management.

♦ The Kenya project failed primarily because of technical institutional, and design problems. The low potential yield of the soil in the project area was not identified at the time of the appraisal. When the design had to be changed

---

[44] OED, Project Performance Audit Report, Sudan: Rahad Irrigation Project, June 13, 1984, p. 59.

[45] OED, World Bank Lending Conditionality, A Review of Cost Recovery in Irrigation Projects, Report 6283, June 25, 1986, p. 30.

accordingly, costs skyrocketed. The inefficiency of the National Irrigation Board, especially its decision-making, was a major problem. Many operational decisions had to be referred to the Board management in Nairobi, but it was far removed from the problem and failed to recognize that an immediate solution was needed. Also, the leadership could not mobilize adequate resources for necessary budgeted items.

- Madagascar's Lake Alaotra project was impeded by numerous technical, engineering, and social problems as a result of project design. Irrigation canals, for example, were dug lower than the rice fields and this caused many subsequent problems. Inadequate technical assessment led to drainage and siltation problems in parts of the project area. Soil studies at appraisal were superficial and thus failed to detect the alkaline soils in the area. As the Impact Evaluation Report notes, the project failed to meet its objective, which was to increase crop diversification. The view of the project area as an intensive rice-growing area was not backed up by historical experience. In fact, the area had long enjoyed a diversified economy. Several social conflicts that were not foreseen at appraisal time developed because of important cultural differences between the original settlers and migrant labor. Even more directly, poor maintenance of the irrigation network, a lack of inputs, poor extension, and a weak implementing authority placed many hurdles in the way of project success.

- The Niena Dionkele Rice Development project in Burkina Faso was a pilot experience and so some problems could be expected. Those that did occur were largely of an institutional nature. In particular, the position and authority of the implementors should have been better defined. There were also problems of recruitment and turnover. A key cooperative specialist, for instance, was never hired.

At the other positive end of the ERR spectrum were Sudan's Rahad project and Cameroon's Second Semry Rice Development project, with re-estimated ERRs of 20 and 16 percent, respectively (PPAR). In Rahad, this increase was due to the higher yields obtained in the absence of long furrow irrigation (which was planned for at reappraisal), the longer time period expected for the life of the assets, and the increase in crop prices in real terms.[46] Project performance in Rahad could have been even better but for early delays due to weaknesses in appraisal, such as technical gaps in project preparation, unrealistic estimates of costs and quantities, and unduly high expectations of organizational arrangements and therefore of implementation effectiveness.

Despite some policy deficiencies that caused marketing problems for Semry rice, the performance of the Second Semry project was good and earned a 16 percent ERR at the time of the PPAR. Excellent management, adoption of the technical package, and the "enclave" nature of the project were considered a key to success. The World Bank's technical supervision also contributed to these accomplishments.

Operational problems that have arisen in these and other SSA projects can be traced to much the same causes. The technical gaps in the Rahad project, for example, have been

---

[46] OED, Sudan: Rahad, PPAR, p. 43.

blamed for cracks that have appeared in some project buildings since the foundations have settled, for cracks in the lining of the discharge basin at the pumping station, and for the recurring siltation of the intake channel. Many buildings in the Bura irrigation settlement project in Kenya have also been threatened by foundation problems.

The importance of the early stages of project appraisal and planning cannot be stressed enough. Soil appraisal in itself is a critical component, as demonstrated in Senegal, where the production of tomatoes had to be abandoned after it was found that the soils were too heavy and saline for profitable yields. This problem should have been recognized before the project was even off the ground. Yet, appraisal crop rotations foresaw tomatoes entering every third year onto paddy fields. As it turned out, the salinity and heavy clay soils of the area forced farmers to use expensive machinery to prepare the land since parcels leveled for paddy flooding had to be converted to sloped ridges for the low irrigation of tomatoes. Proper soil testing should also be accompanied by in-depth mapping to safeguard against such situations.

A significant dimension of project design--and a serious oversight of past projects, as pointed out by some recent irrigation reviews--is the _demand for irrigation services_. Farmers preferences and a decision calculus have to be incorporated into design if irrigation projects are to satisfy the users, encourage their participation, and have good cost recovery and generally sustainable operations.

Conclusion

Judging from the factors that have contributed to the positive and negative performance of irrigation projects in Africa, it appears that a favorable policy environment and good project design and implementation are vital to a successful outcome. Even when policy commitment is high, as in Kenya, design problems--whether because of poor technical assessment, poor engineering design, failure to consider the agricultural and social context, or weak implementing institutions--can adversely affect project performance. It is almost impossible to assign a specific weight to the individual contribution of each of the factors that have had a hand in project failure or success. Nevertheless, this whole range of issues requires careful attention if there is to be any hope of improving agricultural production in Sub-Saharan Africa.

CHAPTER 8: COUNTRY CASE STUDIES

A. PRIVATE SECTOR IRRIGATION IN ZIMBABWE

Johannes ter Vrugt
(Senior Agriculturalist, Southern Africa Department, World Bank)

Zimbabwe has had a long and successful history of private sector irrigation--which should be an encouraging sign to potential investors. Indeed, it has managed to cope fairly well with its cycles of drought, which occur about every five years. Its first major irrigation system was developed between 1910 and 1920 on the Odzani River, where some 600 hectares are still being irrigated. Then in 1920 the British South Africa Company built the Mazoe Dam to irrigate its citrus estates. The dam was raised in 1960 to serve a total of 1,900 hectares of citrus and other crops. By 1950, the country had 7,000 hectares under irrigation, and by 1960 the figure had risen to 20,000 hectares.

Private irrigation expanded even further after 1960, particularly with the establishment of the Sabi-Limpopo Authority in 1965, which brought 40,000 hectares under irrigation for smallholders, estates, and commercial farms. Another significant development was the establishment of the Farmers Irrigation Fund, which provided long-term, low-interest credit for farmers. As a result, commercially irrigated areas jumped from 33,000 hectares in 1965 to 80,000 hectares in 1970. This rapid development can be attributed to the highly successful mix of government involvement and private entrepreneurship. The government built the dams and infrastructure, while the private sector developed the land, built the irrigation networks, and established in-field storage and processing facilities. Thus by 1981, Zimbabwe's total irrigated area was just over 150,000 hectares, and 90 percent of the schemes were private.

In 1985, the National Farm Irrigation Fund (NFIF) was established under the administration of the Agricultural Finance Corporation (AFC) of Zimbabwe. To date, the AFC has financed 156 loans for the development of an additional 7,000 hectares, mainly for winter wheat irrigation, at a total cost of Z$17.4 million (Z$2.25 equals one US dollar). This development brought the total irrigated area to about 165,000 hectares for years with sufficient rainfall (annual rainfall amounts to about 674 millimeters per year in the rain belt).

Both commercial and communal farmers can make use of NFIF, although most of the loans go to the commercial sector (which received 118 out of 156 loans and 98 percent of total investments). However, AFC has also provided other loans for irrigation development since 1981-82, which amounts to about Z$31 million. Most of this support has gone into 50,000 hectares of complementary irrigation facilities needed to give crops a head start before the rains. This brings the total area under irrigation in one form or another to 215,000 hectares.

Present Situation

Another factor that has been a plus for Zimbabwe is its efficient agricultural infrastructure. In addition, its design and construction capability is adequate for the small and medium-scale irrigation provided by some 15 local companies. As a result, Zimbabwe has been charged with overseeing the regional food security objectives of the eight countries of the Southern African Development Coordination Conference (SADCC). Zimbabwe's available water resources and overall use and loss assumption of 10,000 cubic meters per hectare indicate that it is capable of developing about 450,000 hectares of its estimated 600,000 hectares of potentially irrigable land. Further expansion of the irrigation sector has come to a virtual halt in recent months, mainly because of the shortage of foreign exchange required to replace and expand worn-out irrigation equipment and earth-moving equipment, and to procure the necessary spares for electric pumps and engines by the private sector.

Water Supplies

Zimbabwe has three main sources of water supply for irrigation: (1) off the river flow in the summer and some sand-abstraction in the winter; (2) summer flow storage dams (the main source of irrigation in the country); and (3) groundwater supplies tapped by boreholes. Groundwater is o minor importance, however, as sufficient accessible supplies can only be found in two relatively small areas. The water from these sources is often high in calcium and/or magnesium salts and therefore must be used with caution.

Irrigation Systems

Irrigation techniques are well understood in Zimbabwe, in large part because of its efficient agricultural service, which not only assists the user but also interacts closely with the designers of schemes and suppliers of equipment. Most of the area under irrigation, particularly in the commercial farming sector and to a limited extent in the communal farming schemes, is under conventional sprinkler irrigation (about 75 percent). Centerpivot or self-propelled irrigation equipment has not yet been used commercially in the country. However, locally assembled prototypes are being tested for possible manufacture and for sale on the local market and in neighboring countries.

The majority of the communal settlement schemes and sugar estates employ flood irrigation, which is particularly suited to the type of settlement and farming systems there. For deciduous trees and citrus fruit, microjet, undertree sprinkler, and flood systems are all common. In addition, coffee production uses a variety of systems, including microject, hose-and-basin systems, and overhead sprinkler irrigation. With the expansion of horticulture in the past few years, rip system irrigation is also becoming more popular.

Types Of Irrigation Development

Irrigation development in Zimbabwe can be divided into five types: (1) large-scale commercial farms (64 percent); (2) company estates (20 percent); (3) developments overseen by government parastatal bodies, such as the Agricultural and Rural Development Authority (11 percent), (4) small-scale irrigation on commercial lands (3 percent); and (5) small community irrigation schemes (2 percent). A total of 84 percent of all irrigated lands in Zimbabwe are private. The Division of Irrigation in the Department of Agricultural Extension and Technical Services (Agritex) of the Ministry of Lands, Agriculture, and Rural Resettlement (MLARR) is the main agency involved in assisting the communal irrigation sector. The Commercial Farmers Union together with Agritex provides advice for the private sector.

Investment Costs

Stored water for new irrigation schemes is costly. Locally, the estimated cost of contractor-built dams is about Z$4,500 to Z$5,000 per hectare of irrigation. This figure varies, depending on site conditions and the method of construction, whether farmer or contractor-built. The cost of installing irrigation equipment varies greatly, depending on geographical conditions and the type of equipment. The cost of installing hand-moved sprinkler equipment, for example, can range from Z$3,000 to Z$4,000 per hectare.

Irrigated Crops

Zimbabwe is self-sufficient in all food crops except for wheat. It imports about 290,000 tons of wheat a year, which is about one-third of its total annual requirement. Wheat and barley are the main crops grown in the winter, and they are 100 percent irrigated from storage dams from early May to August. Average yields are about 5 tons per hectare. Sugarcane is grown mainly in the Lowlands, in the souther part of Zimbabwe, and is fully irrigated by flood schemes. Yields are good (in excess of 105 tons per hectare) and of high quality owing to the drier weather in this region. A considerable amount of sugar is converted into ethanol for car fuel.

Maize and sorghum have been the main irrigated summer staple crop. Maize yields--at about 8-10 tons per hectare--are among the highest in the world. With the higher rainfall in recent years and the increased production of maize in communal areas, maize reserves have pushed past the strategic reserve level of 2½ million tons. As a result, farmers have been able to shift from irrigated maize into other irrigated crops, such as soybeans and groundnuts. Irrigated cotton production on commercial farms has remained fairly stable, while rainfed communal cotton production has expanded and now accounts for about 50 percent of all cotton production, which was about 350,000 tons in the 1988-89 season. Irrigated cotton yields in the fertile Save Valley are also among the highest in the world (about 4,500 kilograms per hectare).

The Future

In view of the increased yields that have been obtained, the local know-how that has accrued over the years from the use of irrigation, and the ease with which farmers have managed to introduce other irrigated crops, irrigation will undoubtedly continue to expand in Zimbabwe--provided the necessary capital can be provided at reasonable rates of interest.

There are still large areas of land that can be brought under irrigation in all sectors of the farming community, and the country's water resources have not yet been fully developed. Future development depends on the construction of both large and medium-sized multipurpose dams, which are needed to serve the farming community and the urban and industrial sectors.

The main irrigation potential lies in the following areas: Save Valley (90,000 hectares), Lundi River Catchment area (55,000 hectares), Zambezi Valley (12,000 hectares), the Midlands around Kadoma (18,000 hectares), and Mazowe Valley (21,000 hectares). This comes to a total irrigation potential of 196,000 hectares.

B. PRIVATE IRRIGATION DEVELOPMENT IN THE SENEGAL VALLEY

Salah Darghouth
(Principal Engineer, Agriculture Division, Sahelian Countries Department, Africa Region, World Bank)

Many irrigation experts are calling for greater private sector involvement in irrigation development in Sub-Saharan Africa. Up to now, there have been only a few areas in the entire region in which irrigation is being expanded without government help; one of them is the Delta of the Senegal River Valley in Mauritania. This effort is an interesting example for the other countries of the region because it has proved to be quite successful, whereas publicly funded systems have had numerous problems, especially the larger-scale ones. The experience is particularly instructive for the Sahelian countries, especially Senegal, Mauritania, Mali, Niger, Chad, and Burkina Faso.

These countries contain 15 percent of Africa's population, and the demographic pressure on rainfed lands is fast rising. Intermittent droughts over the past decade have also intensified the region's escalating agricultural and food problems. Irrigation offers a promising solution to some of these problems, especially in view of the fact that only about 13 percent of the region's full irrigation potential has been achieved. The central concern of this chapter is what made the Senegal River Valley experience possible, what its limitations have been, and what lessons can be drawn from it.

Background

The Senegal River Basin stretches along the borders of Senegal, Mauritania, and Mali for a distance of 800 kilometers (see Figure 8.1). In Mauritania, the fertile Senegal River Valley in the south gives way to a wide central region of sandy plains and scrub trees. The area to the north is arid and extends into the Sahara. For obvious reasons, population density is highest in the river valley. As a result, crop production has developed on a narrowing resource base. Animal husbandry contributes about 23.4 percent to GDP, in contrast to crop farming, which contributes only 5.4 percent. Fishing has become a substantial source of income during the 1980s (8.3 percent of GDP in the 1980s, particularly in the export-earning sector.

The growth of agricultural production has been severely impaired over the past 20 years by recurrent and worsening droughts. As a result, the 100-millimeter isohyet has moved more than 100 kilometers south, and the Senegal River has reached record low levels. The recent pattern of declining and erratic rainfall has had a severe impact on crops and livestock. The traditional methods of irrigation—flood recession cropping—have also suffered.

In the present circumstances, irrigated production has become an important potential source of growth. Since rainfed production is unlikely to satisfy even half the national demand for cereals, and since flood-recession cropping will decrease as a result of the flood regulation brought about by the Manantali Dam, irrigated agriculture represents the only hope for reducing Mauritania's food deficit.

Present Irrigation Schemes In The Senegal Valley

An area of about 48,000 hectares is already under irrigation on both sides of the Senegal River. Of the numerous scattered schemes found here, 12 can be classified as large-scale operations (averaging 2,000 hectares) and 507 are small-scale operations (averaging 26 hectares). Gravity distribution is practiced after the water has been pumped from the Senegal River. Unlined earthen canals and various control systems are used throughout the valley.

Production at present is devoted mainly to paddy (63 percent), sugarcane (18 percent), maize and sorghum (13 percent), and vegetables (6 percent). Paddy yields are good, averaging 4 tons per hectare. Because of population pressure most of the farms are very small and can barely meet subsistence levels. They average 0.25 hectares where small-scale irrigation is practiced and 1 hectare in areas of large-scale irrigation. These farms satisfy basic food needs, but are too small to generate much cash income and savings.

A turning point in the history of irrigation in this area was the construction of the Diama Dam, commissioned in 1986, and the Manantali Dam, commissioned in 1988. The Diama Dam is designed to prevent seawater intrusion and is expected to increase upstream water levels to 1.5 meters above sea level. The Manantali Dam will have a storage capacity of 11 billion cubic meters and is expected to generate 800 GWH of power. These developments will make it possible to practice double cropping in the full water control areas. Flood recession cropping will be expanded to 50,000 hectares over a transition period of 10-15 years through artificial flood releases from Manantali. The area under irrigation is therefore expected to increase to 375,000 hectares, or almost ten times that currently under full-control irrigation. Despite these recent rapid developments, the area currently under irrigation represents only about 13 percent of the potential.

Constraints On The Development Of Publicly Funded Irrigation

To date, one of the great impediments to irrigation development has been the high investment cost of the irrigated schemes, especially the large ones (US$1 5,000 to 20,000 per hectare) and the difficulty of properly operating and maintaining them. As a result, their viability is not yet ensured. While the large-scale schemes are developed and almost fully managed by the government, the small-scale ones have been developed with public funding, but with villagers' contributions in cash and/or kind. These perimeters cost much less to construct and are easier to manage and maintain, with far better prospects for sustainability, but the number of sites suitable for the construction of such perimeters is limited, and many of the existing ones have encountered management and maintenance problems.

Irrigation development has thus been inhibited by the chronic neglect of operation and maintenance, which is in part due to the low rate of cost recovery. On one hand, farmers have little capacity to pay for use, so that collection rates are low, while on the other hand, the government has limited resources to contribute to irrigation development. Whatever funds it might otherwise allocate to irrigation expansion must now be directed to frequent rehabilitation.

Another constraint is that the world price of rice has been declining for the past decade and thus competition in the form of cheap imported rice has created serious marketing and price constraints on the locally produced price. Still another problem is the lack of incentive, owing to the large government involvement in the supply of inputs and commercial services. Also, the country has no strong and viable bank credit system for farmers. The government has been slow to let the private sector step in and help promote development, but perhaps more important, there has been limited local demand for irrigated crops other than rice. Added to these problems are the lack of local human resources, inadequate legislation on land tenure, lack of coordination among aid donors, design and construction mistakes, and weak extension and research for crops other than paddy. One might well ask how the private scheme was able to do as well as it did.

Recent Expansion Of Privately Funded Irrigation In Mauritania

The irrigated area in Mauritania has tripled from 4,500 hectares in 1984-85 to 15,000 hectares in 1987-88. The publicly funded projects have increased from 4,000 to 6,500 hectares. The most dramatic change has come in privately funded and managed systems, which increased from 500 to 8,500 hectares during the same period, mainly in the Delta area of the Senegal River Valley.

The considerable expansion of irrigation by the private sector over the past three years has been made possible in part by the easy access to land as a result of the 1983 Land Reform Law. Under this law, any Mauritanian individual or collective group can obtain a parcel of land to farm if it is demonstrated that this land is not privately or collectively owned or has not been cultivated by anyone else in the past. In addition, any collective holders of rights who wish to retain collective use of their land are allowed to do so, provided that they organize themselves into a farmers' cooperative. The law also specifies the conditions under which collective concessions can be divided into individual concessions to the members of the cooperative, if they so wish. Traditional crop-sharing systems are abolished under this new structure.

Approximately 35,000 hectares were reportedly distributed under the new law, primarily in the Delta and lower Senegal Valley. By 1988, about 8,500 hectares had been developed—4,000 hectares by cooperatives and 12,000 by individuals. The 1983 law has been found to be comprehensive and equitable. However, land distribution, as conducted so far, has attracted rich people from the north to whom large parcels of land (ranging from 20 to 200 hectares) have been allocated, sometimes hastily, without all the preliminary steps specified in the law.

A second reason for the stepped-up activity in privately supported irrigation is that since the mid-1980s, land values have skyrocketed as a result of recent and contemplated infrastructure investments along the Senegal River undertaken by the Organization pour la Mise en Valeur de Fleuve Senegal (OMVS), a regional organization founded jointly by three riparian countries in 1972. When in 1985 the government issued an executive order to expedite the application of the Land Reform Law, which made it possible to distribute large parcels of land in the Senegal Valley, private investors responded more rapidly than anyone expected. In fact, land speculation has been widespread, by nonresident merchants and businessmen who find profitable investment alternatives to be otherwise scare in the country.

Also important is the fact that the investment required was only about US$800-1,000 per hectare, including the pump, which is much more reasonable than the publicly supported schemes. Moreover, system designs were simple and yields were relatively good (about 3 to 4 tons per hectare).

Another factor contributing to private support is that over the past five years the government has implemented a number of policies aimed at promoting the growth of agriculture. In particular, it announced a progressive adjustment of <u>prices</u> of key agricultural inputs and outputs, which included a 70 percent increase in the consumer price of rice between 1984 and 1988.

There is also a strong motivation to achieve self-sufficiency. Until 1970, Mauritania was self-sufficient in cereals and milk, and was a net exporter of meat. Since then, agricultural output has failed to keep pace with population growth and the country has become increasingly dependent on food imports, particularly in the form of food aid. At present, output meets 40 percent of local consumption needs for milk and 100 percent of meat, with a little surplus left over for export. As for cereals, the country's demand was 300,000 tons in 1988. Food aid covered about 20 percent of this; about 35 percent was met through commercial imports, and only 45 percent was produced locally.

Despite its impressive performance, private sector irrigation in Mauritania has not been without problems. For one thing, the service market has been both narrow and weak. Technical problems have also been numerous: they range from the lack of basic technical support and inputs, to physical problems associated with the site location in the delta, where drainage is poor and the water table is shallow. Also, until recently, the government has not been fully committed to improving the price markets for producers, who face an added risk in the lack of credit, particularly for the small holders. An additional problem is the social tensions caused by the land reforms, which have been difficult to comprehend by many peasant and village groups that have taken customary rights for granted for many generations.

Prospects For Future Irrigation Development In Mauritania

The potential for economic growth in Mauritania depends in large measure on irrigated production. Other sectors of the economy that were previously engines of growth now appear to be slowing down. This is the case for iron ore which is dependent on industrial growth in the rest of the world; fishing, which is constrained by the sustainable level of yearly catch; and livestock, where excessive expansion could accelerate desertification. In contrast, irrigation has a large potential for expansion. Since rainfed production is unlikely to satisfy even half the national demand for cereals, and since flood-recession cropping will decrease as a result of flood regulation brought about by the Manantali dam, irrigated agriculture today represents the only possibility for reducing the country's food deficit.

An effort is now under way to find a solution to the various problems that have plagued irrigation development in Mauritania as well as the Senegal River Valley and to achieve more sustained private sector promotion. One of the initial suggestions put forth was that costs be held down by adopting medium-scale schemes. These would be less expensive, and would ensure greater farmer participation, more rapid construction, and more efficient use of the government's scarce resources than the scattered small-scale schemes currently in use. It has also been proposed that private farmers be allowed to develop schemes at their own expense. Benefits could be increased by setting schemes up in areas where the soil would allow crops other than paddy to be cultivated. Eventually, private commercial farms and agrobusiness could move into more remunerative crops than paddy. In addition, cropping intensity could be increased in the existing small subsistence farms. Some have also suggested that benefits could be derived from more in-depth economic analysis of irrigation projects. More work could be done, for example, on valuing the opportunity cost of labor and the impact of irrigation on the generation of labor and on immigration to cities, the creation of new activities such as continental fisheries, and so on.

Operation and maintenance could be improved if steps were taken to facilitate farmers' participation. Costs could be kept down by consolidating government development agencies and the training staff in charge of operation and maintenance. A more effective cost recovery policy should also be given some thought. In addition, government involvement in the provision of inputs and commercial services to farmers could be reduced. Some progress has already been achieved in this respect, but proper reforms should be set up to generate greater private sector involvement so that it could begin while the government is slowly withdrawing from some of these activities. One problem that will persist in the short to medium term, however, is the limited local demand for products other than paddy, although proper channels for export marketing might be investigated as a possible solution.

C. PRIVATE SECTOR SMALL-SCALE IRRIGATION IN NIGERIA

Lewis G. Campbell
(Senior Agriculturalist, Western Africa Department, World Bank)

Small-scale irrigation has been practiced in Nigeria since precolonial times, but the methods used vary from one area to another. Nigeria can be divided into three basic zones according to the rainfall received. The northern zone receives less than 700 millimeters of rain spread over about 140 days, and the timing of the rainy season is often erratic. Traditional short-season crops such as sorghum, millet, groundnut, and cotton are grown there and planting coincides by and large with the rainy months. In the middle belt, the annual rainfall may reach 1,000 millimeters and the rainy season may extend to about 180 days, but short-term crops often suffer periodic droughts and many fields lie fallow for more than half the year. Although the southern part of the country receives more rainfall, short-term crops will not grow there during the three to four months of the dry season without irrigation.

In the drier areas with reliable water resources, farmers generally use the simple shadouf or rope-and-bucket system for raising water from rivers or shallow wells. This method is used primarily for growing vegetables. Rice is grown in the wetter valley bottoms, where farmers can use simple techniques to control the level of runoff. During the petroleum boom of the late 1950s, the area under wet season rice was expanded, but a substantial part of this production still depended on informal methods of irrigation, which were undertaken at the initiative of the farmers or landowners and without significant, if any, government assistance. It has only been since the late 1960s that the government has become involved in irrigation, primarily in formal schemes. The total area covered by these schemes has expanded from about 5,000 hectares in 1968 to 30,000 in 1988. The cost of these schemes has averaged in excess of $30,000 per hectare.

Since the early 1980s, the World Bank has supported agricultural development projects (ADPs) in the northern states that have promoted the use of shallow groundwater resources in the lower-lying areas (fadamas) for small-scale irrigation. Farmer response to these endeavors has been far beyond that originally expected. The country's recent experience in the development of small-scale irrigation—particularly in the northern states of Sokoto, Kano, and Bauchi—provides an encouraging example to would-be investors.

Irrigation Development

In the mid-1970s, Nigeria received assistance from the World Bank to address a number of problems in the declining agricultural sector. The country's agricultural policy at that time was geared to increasing local food production, which was not keeping up with demand. Three enclave ADPs (Gusau, Funtua, and Gombe) were established in the states of Sokoto, Kaduna, and Bauchi to help farmers apply improved technologies in order to increase grain production. The strategy focused on improved varieties, good-quality seed, fertilizers, crop protection, and building a basic rural infrastructure. These early projects were fairly successful, and on the strength of their results Nigeria secured further assistance for expanding the programs throughout each of the three states and into another state, Kano.

The government carefully reviewed the earlier projects, determined to apply the lessons learned to the new ones.

In many cases where yields had not increased, for example, water deficiency was to blame, and so the next generation of ADPs sought to control the critical soil moisture regimes, both on the surface and in the ground. Groundwater, which for generations had supplied the shallow wells and shadoufs, received particular attention. Groundwater surveys showed that considerable potential existed for abstracting water relatively inexpensively from the high-yielding shallow aquifers in the low-lying fadama areas. The groundwater potential in the northern states was estimated to be 280,000 hectares in Sokoto, 236,350 hectares in Bauchi, 35,000 hectares in Kano, 80,000 hectares in Kaduna, and 46,000 hectares in Katsina for a total potential of 677,350 hectares.

The ADPs in Sokoto, Kano, and Bauchi organized shallow well-drilling services and taught farmers how to use shallow well water for irrigation. The projects concentrated on abstracting water from existing dug wells with greater efficiency by replacing the shadoufs and rope and buckets with small (3 to 5 horsepower) portable centrifugal pumps driven by gas engines. Low-cost methods were also used to establish shallow wells in the fadamas and dry river beds. The ADPs examined the successful shallow tubewell operations in other countries, particularly in Asia, and tried to adopt the simpler technologies for establishing them. They now use one of three methods, depending on the environment: (1) rotary drilling, (2) percussion bailer, or (3) jetting or washboring.

Rotary drilling is the most expensive of the three methods, but is generally more rapid than the others if the operating crews are well organized. However, special transport facilities are required to move the equipment between locations and this pushes up the cost. The percussion bailer is simple and inexpensive to use. It can be manufactured by local machine shops and is not subject to serious mechanical failures. It is easily managed by relatively unskilled labor. The jetting and washboring method is best applied in light soils of the type found in or near riverbeds. The technique is simple but requires adequate water to serve as the primary excavation and spoil-transporting medium. Each ADP operates about three rigs to provide tubewell drilling. Farmer interest in irrigated farming has escalated since the new and inexpensive well and pumping equipment was installed. The ADPs subsidize about 50 percent of the cost of establishing the well, and the pumping equipment is made available at the actual purchase price plus the cost of handling. Table 8.1 shows the wells established in Sokoto, Kano, and Bauchi since 1983, when the ADP programs were introduced.

Table 8.1

THE COST OF ESTABLISHING WELLS IN NORTHERN NIGERIA

STATE	Total successful tubewells at end of 1988	Unit Costs (N)	
		MECHANICAL RIGS	WASHBORES
Sokoto	2,200	919.00	808.00
Kano	4,000	1,563.00	405.00
Bauchi	2,450	1,088.00	532.00

Present Situation

As already mentioned, farmer demand for the tubewell service has exceeded all expectations, and it will be difficult to clear the backlog unless the number of operating units is substantial increased. The success rate of shallow tubewells has jumped from less than 50 percent at the start of the program to more than 60 percent. The discharge rate of wells has ranged from 2,000 to 5,000 gallons per hour. Any well that tested below 3,000 gallons per hour in the dry season was not regarded successful as an economical unit. The light, gasoline-driven pumps were bought from the ADP Commercial Services Company at a cost of US$400 to $600, which most farmers found affordable before the devaluation of the local currency.

These wells are capable of irrigating up to 2 hectares of land, depending on the cropping pattern, and at least two short-term crops can be produced each year. In practice, farmers do not generally grow more than half a hectare of irrigated crops, unless they are able to secure additional labor and the markets for the perishable crops do not become saturated. The major market for the perishable crops is in the cities of the south, notably Lagos and Ibadan. If the logistics of handling, transport, and marketing were not properly organized, the returns to the farmer could be considerably less.

The merits of irrigation are well understood by the farmers, and the low cost of the initial investment relative to the returns has helped to encourage them to adopt the practice. They have found the exploitation of the shallow aquifers particularly attractive as they are able to make the investment without depending on the government, or having it involved. They are responsible for operating the system and for scheduling its use according to their needs, and to other activities. Minimal infrastructure is required to develop the irrigation system and the costs are easily justified and generally within the financing capacity of the farmers.

The system is simple and can be easily managed by even the least skilled farmers. The portability of the pumping equipment provides the security needed to safeguard against loss or damage, which often occurs when such equipment is left unattended in the field. The single-cylinder gasoline engines are in common use in the rural areas and village mechanics are skilled in their repair and maintenance. The operational characteristics of the centrifugal pump are easily taught to farmers and mechanics, and they can be used and maintained without much trouble.

Issues Requiring Attention

Although great progress has been made in introducing this simple tubewell technology for small-scale irrigation in northern Nigeria, the rate of development is still far short of its potential. The expansion rate could be increased considerably if some of the present constraints could be overcome. Effort should also be put into increasing the efficiency of irrigation facilities, maximizing the returns to investment, and ensuring that the schemes are sustainable and that the water resources are safely managed. The following points in particular merit close attention.

- Well-drilling service. The efforts of the ADPs to provide well drilling services are commendable, but they should be regarded only as catalysts for a larger program in which the private sector plays a large role. At the current rate at which successful wells are being established, it would take more than 50 years to exploit the irrigation potential of shallow groundwater. ADP tubewell drilling facilities should not be expanded, however, because the government-run services are inadequate in several respects. Considerable scope exists for increasing the performance of the present drilling rigs, including their success rate, which has been lower than desired because of local pressure to establish wells in marginal areas. The substantial subsidy provided by the project has also tended to encourage waste in the use of the service. These weaknesses would be rectified if the private sector could organize and offer a reliable service. The ADPs could then concentrate on promoting private drilling and on training village artisans, mechanics, or small contractors to maintain the services. Those who successfully developed the competence could be assisted in raising the capital, which is not prohibitive, needed to establish reliable water wells and maintenance services.

- ◆ Organizing credit. Before the devaluation of the local currency, farmers had no difficulty finding the funds from their savings to meet the cost of wells and pumping equipment. With the recent sevenfold plus devaluation, the cost of imported pumps and engines has increased substantially—from about N500.00 to N3,500.00 each for pumps and from about N1,000.00 to N2,500.00 for wells. Consequently, few farmers are now able to make the full payment at the time of installation. Because individual loans and the level of debt service would be small, commercial banks would probably shy away from such loans and tend to favor more secure and rewarding operations in the commercial sector. An alternative credit mechanism could be group or cooperative borrowing from commercial banks, using rural cooperative credit organizations where they already exist. The ADs could play a useful role in promoting the organization of such group borrowing and repayment arrangements.

- ◆ Improving marketing arrangements. There is substantial waste of perishable commodities produced for the local market owing to improper handling and transport, as well as delays in getting the produce to the consumer. Better techniques should be introduced for protecting, harvesting, crating, and transporting these crops to the markets. Producers should also explore alternative markets, such as processing plants and the export markets, particularly in the neighboring states as well as in Europe. Because the potential for irrigated grain crops at the current prices and the present demand appear highly favorable, these crops could be included in appropriate rotations with other short-term crops.

- ◆ Irrigation efficiency. Farmers have not even begun to realize the potential of their irrigation facilities. Most are satisfied with yield increases of 25-40 percent in the wet months and a second crop in the dry months, which, admittedly, brings them considerably more income. However, the standard of agronomic management is still low, and the benefits of irrigation investments could be increased substantially. Institutional arrangements should therefore be made for maximizing the benefits and transferring them to farmers. For example, varieties could be improved, fertilizer use could be increased to optimum balanced levels, and crops protection could be improved. More effective techniques for using the available water should also be sought.

- ◆ Aquifer management. Aquifers should not be overpumped, particularly in the smaller fadama areas. The government should be responsible for monitoring the extraction of water and for overseeing the preservation of aquifers. However, it should not impose such complicated regulations that farmers would become discouraged from investing in irrigation and be tempted to short-circuit or undermine the system and defeat the basic objective—which is to conserve the resource.

- **Local manufacture of pumps.** To avoid placing a further burden on the country's limited foreign exchange resources, an effort should be made to manufacture much of the simple equipment locally, such as self-priming low head centrifugal pumps for small-scale irrigation. The long-term need for such pumps could well justify installing the equipment required to manufacture them. The prospects for such manufacturing should be explored.

Conclusion

Although Nigeria should not ignore the substantial investment it has already made in large-scale irrigation projects, many of which are yet to be completed, it should put more emphasis on small-scale irrigation. The investment level is relatively low and the costs can be borne directly by farmers. However, some short- to medium-term credit mechanisms will have to be introduced before small farmers will be able to undertake such investments. The technology is simple and can be handled by Nigerian farmers and artisans, whether they will be operating or repairing and maintaining the equipment and facilities. These factors, in the presence of Nigeria's good economic climate for the crop commodities, constitute an optimistic forecast for the sustainability of irrigated agriculture in this part of Sub-Saharan Africa.

Distributors of World Bank Publications

ARGENTINA
Carlos Hirsch, SRL
Galeria Guemes
Florida 165, 4th Floor-Ofc. 453/465
1333 Buenos Aires

AUSTRALIA, PAPUA NEW GUINEA, FIJI, SOLOMON ISLANDS, VANUATU, AND WESTERN SAMOA
D.A. Books & Journals
11-13 Station Street
Mitcham 3132
Victoria

AUSTRIA
Gerold and Co.
Graben 31
A-1011 Wien

BAHRAIN
Bahrain Research and Consultancy Associates Ltd.
P.O. Box 22103
Manama Town 317

BANGLADESH
Micro Industries Development Assistance Society (MIDAS)
House 5, Road 16
Dhanmondi R/Area
Dhaka 1209

Branch offices:
156, Nur Ahmed Sarak
Chittagong 4000

76, K.D.A. Avenue
Kulna

BELGIUM
Publications des Nations Unies
Av. du Roi 202
1060 Brussels

BRAZIL
Publicacoes Tecnicas Internacionais Ltda.
Rua Peixoto Gomide, 209
01409 Sao Paulo, SP

CANADA
Le Diffuseur
C.P. 85, 1501B rue Ampère
Boucherville, Quebec
J4B 5E6

CHINA
China Financial & Economic Publishing House
8, Da Fo Si Dong Jie
Beijing

COLOMBIA
Enlace Ltda.
Apartado Aereo 34270
Bogota D.E.

COSTA RICA
Libreria Trejos
Calle 11-13
Av. Fernandez Guell
San Jose

COTE D'IVOIRE
Centre d'Edition et de Diffusion Africaines (CEDA)
04 B.P. 541
Abidjan 04 Plateau

CYPRUS
MEMRB Information Services
P.O. Box 2098
Nicosia

DENMARK
SamfundsLitteratur
Rosenoerns Allé 11
DK-1970 Frederiksberg C

DOMINICAN REPUBLIC
Editora Taller, C. por A.
Restauracion e Isabel la Catolica 309
Apartado Postal 2190
Santo Domingo

EL SALVADOR
Fusades
Avenida Manuel Enrique Araujo #3530
Edificio SISA, 1er. Piso
San Salvador

EGYPT, ARAB REPUBLIC OF
Al Ahram
Al Galaa Street
Cairo

The Middle East Observer
8 Chawarbi Street
Cairo

FINLAND
Akateeminen Kirjakauppa
P.O. Box 128
SF-00101
Helsinki 10

FRANCE
World Bank Publications
66, avenue d'Iéna
75116 Paris

GERMANY, FEDERAL REPUBLIC OF
UNO-Verlag
Poppelsdorfer Allee 55
D-5300 Bonn 1

GREECE
KEME
24, Ippodamou Street Platia Plastiras
Athens-11635

GUATEMALA
Librerias Piedra Santa
Centro Cultural Piedra Santa
11 calle 6-50 zona 1
Guatemala City

HONG KONG, MACAO
Asia 2000 Ltd.
Mongkok Post Office
Bute Street No. 37
Mongkok, Kowloon
Hong Kong

HUNGARY
Kultura
P.O. Box 139
1389 Budapest 62

INDIA
Allied Publishers Private Ltd.
751 Mount Road
Madras - 600 002

Branch offices:
15 J.N. Heredia Marg
Ballard Estate
Bombay - 400 038

13/14 Asaf Ali Road
New Delhi - 110 002

17 Chittaranjan Avenue
Calcutta - 700 072

Jayadeva Hostel Building
5th Main Road Gandhinagar
Bangalore - 560 009

3-5-1129 Kachiguda Cross Road
Hyderabad - 500 027

Prarthana Flats, 2nd Floor
Near Thakore Baug, Navrangpura
Ahmedabad - 380 009

Patiala House
16-A Ashok Marg
Lucknow - 226 001

INDONESIA
Pt. Indira Limited
Jl. Sam Ratulangi 37
P.O. Box 181
Jakarta Pusat

IRELAND
TDC Publishers
12 North Frederick Street
Dublin 1

ITALY
Licosa Commissionaria Sansoni SPA
Via Benedetto Fortini, 120/10
Casella Postale 552
50125 Florence

JAPAN
Eastern Book Service
37-3, Hongo 3-Chome, Bunkyo-ku 113
Tokyo

KENYA
Africa Book Service (E.A.) Ltd.
P.O. Box 45245
Nairobi

KOREA, REPUBLIC OF
Pan Korea Book Corporation
P.O. Box 101, Kwangwhamun
Seoul

KUWAIT
MEMRB Information Services
P.O. Box 5465

MALAYSIA
University of Malaya Cooperative Bookshop, Limited
P.O. Box 1127, Jalan Pantai Baru
Kuala Lumpur

MEXICO
INFOTEC
Apartado Postal 22-860
14060 Tlalpan, Mexico D.F.

MOROCCO
Societe d'Etudes Marketing Marocaine
12 rue Mozart, Bd. d'Anfa
Casablanca

NETHERLANDS
InOr-Publikaties b.v.
P.O. Box 14
7240 BA Lochem

NEW ZEALAND
Hills Library and Information Service
Private Bag
New Market
Auckland

NIGERIA
University Press Limited
Three Crowns Building Jericho
Private Mail Bag 5095
Ibadan

NORWAY
Narvesen Information Center
Bertrand Narvesens vei 2
P.O. Box 6125 Etterstad
N-0602 Oslo 6

OMAN
MEMRB Information Services
P.O. Box 1613, Seeb Airport
Muscat

PAKISTAN
Mirza Book Agency
65, Shahrah-e-Quaid-e-Azam
P.O. Box No. 729
Lahore 3

PERU
Editorial Desarrollo SA
Apartado 3824
Lima

PHILIPPINES
National Book Store
701 Rizal Avenue
P.O. Box 1934
Metro Manila

POLAND
ORPAN
Patac Kultury i Nauki
00-901 Warszawa

PORTUGAL
Livraria Portugal
Rua Do Carmo 70-74
1200 Lisbon

SAUDI ARABIA, QATAR
Jarir Book Store
P.O. Box 3196
Riyadh 11471

MEMRB Information Services
Branch offices:
Al Alsa Street
Al Dahna Center
First Floor
P.O. Box 7188
Riyadh

Haji Abdullah Alireza Building
King Khaled Street
P.O. Box 3969
Damman

33, Mohammed Hassan Awad Street
P.O. Box 5978
Jeddah

SINGAPORE, TAIWAN, MYANMAR, BRUNEI
Information Publications Private, Ltd.
02-06 1st Fl., Pei-Fu Industrial Bldg.
24 New Industrial Road
Singapore 1953

SOUTH AFRICA, BOTSWANA
For single titles:
Oxford University Press Southern Africa
P.O. Box 1141
Cape Town 8000

For subscription orders:
International Subscription Service
P.O. Box 41095
Craighall
Johannesburg 2024

SPAIN
Mundi-Prensa Libros, S.A.
Castello 37
28001 Madrid

Libreria Internacional AEDOS
Consell de Cent, 391
08009 Barcelona

SRI LANKA AND THE MALDIVES
Lake House Bookshop
P.O. Box 244
100, Sir Chittampalam A. Gardiner Mawatha
Colombo 2

SWEDEN
For single titles:
Fritzes Fackboksforetaget
Regeringsgatan 12, Box 16356
S-103 27 Stockholm

For subscription orders:
Wennergren-Williams AB
Box 30004
S-104 25 Stockholm

SWITZERLAND
For single titles:
Librairie Payot
6, rue Grenus
Case postal 381
CH 1211 Geneva 11

For subscription orders:
Librairie Payot
Service des Abonnements
Case postal 3312
CH 1002 Lausanne

TANZANIA
Oxford University Press
P.O. Box 5299
Dar es Salaam

THAILAND
Central Department Store
306 Silom Road
Bangkok

TRINIDAD & TOBAGO, ANTIGUA BARBUDA, BARBADOS, DOMINICA, GRENADA, GUYANA, JAMAICA, MONTSERRAT, ST. KITTS & NEVIS, ST. LUCIA, ST. VINCENT & GRENADINES
Systematics Studies Unit
#9 Watts Street
Curepe
Trinidad, West Indies

TURKEY
Haset Kitapevi, A.S.
Istiklal Caddesi No. 469
Beyoglu
Istanbul

UGANDA
Uganda Bookshop
P.O. Box 7145
Kampala

UNITED ARAB EMIRATES
MEMRB Gulf Co.
P.O. Box 6097
Sharjah

UNITED KINGDOM
Microinfo Ltd.
P.O. Box 3
Alton, Hampshire GU34 2PG
England

URUGUAY
Instituto Nacional del Libro
San Jose 1116
Montevideo

VENEZUELA
Libreria del Este
Aptdo. 60.337
Caracas 1060-A

YUGOSLAVIA
Jugoslovenska Knjiga
YU-11000 Belgrade Trg Republike

RECENT WORLD BANK TECHNICAL PAPERS *(continued)*

No. 90 Candoy-Sekse, *Techniques of Privatization of State-Owned Enterprises,* vol. III: *Inventory of Country Experience and Reference Materials*

No. 91 Reij, Mulder, and Begemann, *Water Harvesting for Plant Production: A Comprehensive Review of the Literature*

No. 92 The Petroleum Finance Company, Ltd., *World Petroleum Markets: A Framework for Reliable Projections*

No. 93 Batstone, Smith, and Wilson, *The Safe Disposal of Hazardous Wastes: The Special Needs and Problems of Developing Countries*

No. 94 Le Moigne, Barghouti, and Plusquellec, *Technological and Institutional Innovation in Irrigation*

No. 95 Swanson and Wolde-Semait, *Africa's Public Enterprise Sector and Evidence of Reforms*

No. 96 Razavi, *The New Era of Petroleum Trading: Spot Oil, Spot-Related Contracts, and Futures Markets*

No. 97 Asia Technical Department and Europe, Middle East, and North Africa Technical Department, *Improving the Supply of Fertilizers to Developing Countries: A Summary of the World Bank's Experience*

No. 98 Moreno and Fallen Bailey, *Alternative Transport Fuels from Natural Gas*

No. 99 International Commission on Irrigation and Drainage, *Planning the Management, Operation, and Maintenance of Irrigation and Drainage Systems: A Guide for the Preparation of Strategies and Manuals* (also in French, 99F)

No. 100 Veldkamp, *Recommended Practices for Testing Water-Pumping Windmills*

No. 101 van Meel and Smulders, *Wind Pumping: A Handbook*

No. 102 Berg and Brems, *A Case for Promoting Breastfeeding in Projects to Limit Fertility*

No. 103 Banerjee, *Shrubs in Tropical Forest Ecosystems: Examples from India*

No. 104 Schware, *The World Software Industry and Software Engineering: Opportunities and Constraints for Newly Industrialized Economies*

No. 105 Pasha and McGarry, *Rural Water Supply and Sanitation in Pakistan: Lessons from Experience*

No. 106 Pinto and Besant-Jones, *Demand and Netback Values for Gas in Electricity*

No. 107 Electric Power Research Institute and EMENA, *The Current State of Atmospheric Fluidized-Bed Combustion Technology*

No. 108 Falloux, *Land Information and Remote Sensing for Renewable Resource Management in Sub-Saharan Africa: A Demand-Driven Approach* (also in French, 108F)

No. 109 Carr, *Technology for Small-Scale Farmers in Sub-Saharan Africa: Experience with Food Crop Production in Five Major Ecological Zones*

No. 110 Dixon, Talbot, and Le Moigne, *Dams and the Environment: Considerations in World Bank Projects*

No. 111 Jeffcoate and Pond, *Large Water Meters: Guidelines for Selection, Testing, and Maintenance*

No. 112 Cook and Grut, *Agroforestry in Sub-Saharan Africa: A Farmer's Perspective*

No. 113 Vergara and Babelon, *The Petrochemical Industry in Developing Asia: A Review of the Current Situation and Prospects for Development in the 1990s*

No. 114 McGuire and Popkins, *Helping Women Improve Nutrition in the Developing World: Beating the Zero Sum Game*

No. 115 Le Moigne, Plusquellec, and Barghouti, *Dam Safety and the Environment*

No. 116 Nelson, *Dryland Management: The "Desertification" Problem*

No. 117 Barghouti, Timmer, and Siegel, *Rural Diversification: Lessons from East Asia*

No. 118 Pritchard, *Lending by the World Bank for Agricultural Research: A Review of the Years 1981 through 1987*

No. 119 Asia Energy Technical Department, *Flood Control in Bangladesh: A Plan for Action*

No. 120 Plusquellec, *The Gezira Irrigation Scheme in Sudan: Objectives, Design, and Performance*

No. 121 Listorti, *Environmental Health Components for Water Supply, Sanitation, and Urban Projects*

No. 122 Dessing, *Support for Microenterprises: Lessons for Sub-Saharan Africa*

The World Bank

Headquarters
1818 H. Street, N.W.
Washington, D.C. 20433, U.S.A.

Telephone: (202) 477-1234
Facsimile: (202) 477-6391
Telex: WUI 64145 WORLDBANK
       RCA 248423 WORLDBK
Cable Address: INTBAFRAD
               WASHINGTONDC

European Office
66, avenue d'Iéna
75116 Paris, France

Telephone: (1) 40.69.30.00
Facsimile: (1) 47.20.19.66
Telex: 842-620628

Tokyo Office
Kokusai Building
1-1 Marunouchi 3-chome
Chiyoda-ku, Tokyo 100, Japan

Telephone: (3) 214-5001
Facsimile: (3) 214-3657
Telex: 781-26838

Cover design by Walton Rosenquist

ISBN 0-8213-1554-4